Microbial Ecology

Basic Microbiology

EDITOR: J. F. WILKINSON

Volume 5
Microbial Ecology

R. CAMPBELL
BSc, MS, PhD
Department of Botany
University of Bristol

A HALSTED PRESS BOOK

JOHN WILEY & SONS
NEW YORK – TORONTO

Osney Mead, Oxford, OX2 0EL,
8 John Street, London, WC1N 2ES,
9 Forrest Road, Edinburgh, EH1 2QH,
P.O. Box 9, North Balwyn, Victoria, Australia.

First published 1977

Published in the U.S.A. by
Halsted Press, a Division of
John Wiley & Sons, Inc.,
New York.

Library of Congress Cataloging in Publication Data

Campbell, Richard Ewen.
Microbial ecology.

(Basic microbiology; v. 5)
Includes index.
1. Microbial ecology I Title.
QR100.C35 576'.15 77-5460
ISBN 0-470-99164-X

Printed in Great Britain

Contents

	Preface	vii
1	Introduction and methods of study	1
2	Concepts in microbial ecology	9
3	Microbial conversions of carbon in the environment	17
4	Microbial conversions of nitrogen in the environment	34
5	Microbial conversions of other elements in the environment	51
6	The structure and dynamics of microbial populations in soil	63
7	The structure and dynamics of microbial populations in water	92
8	The structure and dynamics of microbial populations in the air	128
9	Conclusions	142
	Index	143

Preface

This book is an attempt to describe the activities and the distribution of micro-organisms on the basis of both the chemical transformations that they mediate and the environments in which they live. The physiological approach to microbial ecology has not been stressed because I think that there is a need to consider the subject in terms of communities of interacting organisms. This approach is hampered by the very uneven quality and quantity of information on the various groups of micro-organisms and by the difference in emphasis made by the specialists who study each group. I have tried to give a balanced account which reflects the relative importance of algae, protozoa, bacteria and fungi in the different habitats.

I thank all my colleagues in the Departments of Botany and Bacteriology, Bristol University, who have endured my questions and requests for help. In particular Professor D.C. Smith, Dr A. Beckett and S. Shales gave much advice during the preparation of the manuscript. I would also like to thank Tim Colborn who helped with many of the illustrations and Jean Hancock for all the typing. Finally I would like to acknowledge all those whose papers and books have supplied ideas or information but who have not been specifically quoted in the text.

1 Introduction and methods of study

INTRODUCTION

Ecology is the study of the inter-relations between organisms in the environments where they live. "Natural" environments are usually considered, but also within the scope of ecology are the environments of cities and industrial complexes, indeed in most industrialized countries there are no areas left unaffected by man. Furthermore man is often most interested in severely disturbed areas: he wants to know what happens to the organisms living in the soil when he practices intensive agriculture or how rivers cope with the effluents he pours into them. The recent interest in ecology has prompted research along two main lines, firstly to find out how the natural environments operate while there is yet time, and secondly to study the influence of man on these normal processes.

The ecology of micro-organisms includes that of bacteria, fungi, protozoa and algae (further reading 8, for an outline of these groups), which range in size from less than a micrometre (μm) to at most a few tens of μm, and show a great diversity in their requirements for life and their tolerance of unfavourable conditions. They may be aerobic or anaerobic, heterotrophic or chemo- or photo-autotrophic (further reading 2 and 8). The basic principles and concepts of the ecology of micro-organisms are similar to those already worked out for higher plants and animals, but micro-organisms also present additional problems and difficulties. For example they are too small to see with the naked eye; you cannot just go into a field and count the bacteria as you would go and survey the plants by identifying and measuring them. One of the major problems in microbial ecology is therefore methodology, and the resulting scarcity of reliable basic data on the distribution, numbers and productivity of micro-organisms remains one of the restraints on the development of microbial ecology as a quantitative science. We will consider what methods are available later in this chapter.

The small size of micro-organisms results in very intimate contact with their environment and they have a very high surface area to volume ratio; a large surface through which the environmental factors may act on the organism. Many of the heterotrophs also have extracellular enzymes which are not contained within the buffered and ionically controlled environment of the cell. Micro-organisms are therefore particularly sensitive to changes in the levels of temperature, light, pH, organic and inorganic nutrients, carbon dioxide, oxygen, water etc., and extreme levels may well have more effect on micro-organisms than they do on higher plants and animals which are in some ways more isolated from the environment (further reading 2 for a detailed consideration of the effect of environment on micro-organisms).

There are many different habitats on a macroscopic scale which have well recognized, and often characteristic, communities of higher plants and animals. It is probable that their microbial flora and fauna is also characteristic. Within these habitats there are microhabitats which for example may have a characteristic soil, but for microbes it will matter which part of that soil (which horizon, Chapter 6) is considered and even which soil crumb; whether it be mineral or organic matter for example will affect the nutrients available. In microbiology habitats of micrometre, or at the most millimetre, dimensions are important. You must adjust the scale of your thoughts and imagination to encompass both the large habitat variations and also the very small dimensions over which variations in microbial communities can occur.

It follows from the great diversity of morphology and chemical ability found in micro-organisms that they can live, or at least survive, in a remarkable variety of habitats. They are present everywhere on earth where higher plants and animals exist and also in much less hospitable places such as the deep ocean sediments or the upper layers of the atmosphere. Some can survive the conditions in space and they have been transported to the moon by man and can survive there outside spacecraft for at least many months.

Micro-organisms are generally so well dispersed that any that are capable of living and growing in a particular environment will most probably be there. Conversely, if a micro-organism is not there then there is usually a good reason. It is therefore difficult to introduce a micro-organism into an established environment where it does not at present exist. Any disturbance of the existing balance of biotic and abiotic factors may cause a temporary shift in the population but a new equilibrium will eventually be established. Communities, including microbial ones, are inherently stable but at the same time they are dynamic structures, and we will see examples of this apparent contradiction in later chapters.

METHODS OF STUDY IN MICROBIAL ECOLOGY

It is not the object of this section to give detailed methods, these are readily available elsewhere (further reading 1, 5, 6, and 7), but rather to set out some principles and general methods.

The micro-organisms have to be observed with the microscope or made visible by culturing them into macroscopic colonies. Both these processes are time consuming and the amount of material studied is usually very small in relation to the whole habitat, but the heterogeneity of microbial communities requires that to get representative data the sample should, in fact, be rather large. This is a basic conflict in methodology of microbial ecology. In the past microbial ecologists have often ignored the statistical approach, which is considered essential in normal ecology but this is now changing and it is realized that a few hard-won figures whose degree of precision can be estimated are more valuable than large amounts of data which cannot be objectively compared or analysed. No matter what particular method you use it is going to be a long, and probably tedious, job to get reliable data which express the relationships between different micro-organisms or between them and their environment.

The most common methods count the numbers in some way, but this does not give information on whether the organisms were active or dormant and takes no account of size, so it is not possible to estimate the ecological significance of an organism from its numbers alone. The biomass in the environment tells more than numbers, but does not take into account the differing rates of metabolism in different organisms. Generally the metabolic activity per unit mass decreases as the mass increases. It can be that a single, large, protozoon is equivalent to a single, much smaller, more metabolically active bacterium (Table 1.1). It is easier to measure numbers than biomass and it is most difficult of all to measure metabolic activity. Each of these methods will now be considered.

Table 1.1 Approximate oxygen uptake (Q_{O_2}), cell size and the oxygen consumed per cell under laboratory conditions: note that single organisms may contribute very different amounts to the total respiration in the environment. The oxygen consumption of *Azotobacter* is exceptionally high because of the nitrogen fixing ability of this bacterium (Chapter 4).

Organism	Q_{O_2}*	Wet weight of a single cell mg	μl oxygen used per cell per hour**
BACTERIA			
Azotobacter chroococcum	6000	2.2×10^{-9}	2.6×10^{-4}
Pseudomonas fluorescens	60	1.7×10^{-10}	2.0×10^{-3}
FUNGI			
Saccharomyces cerviseae	95	7×10^{-8}	1.3×10^{-6}
PROTOZOA			
Paramecium aurelia	6	5×10^{-4}	6.2×10^{-4}
ALGAE			
Chlorella vulgaris	70	3×10^{-8}	4×10^{-7}

* μl of oxygen/mg dry weight/hour
** assumes a water content of 80% by weight

Direct examination involves the observation of the organisms, usually with the light microscope, and counting their numbers or measuring the length of filamentous organisms like fungi and some algae. The results (direct counts) can be expressed as numbers per unit area, volume or weight, and the size may be measured to give an estimate of biomass. There has recently been a great increase in the variety of techniques used, including phase, Nomarski interference and fluorescent microscopy. The latter may involve general staining of organisms e.g. with acridine orange, or may be made very specific for a particular organism by linking the fluorescent stain with an antibody. Electron microscopes, both transmission and scanning, have also been used for direct examination of micro-organisms (further reading 6).

There is a wide variety of methods involving the culturing of micro-organisms. It may be necessary to dilute or concentrate the sample so that a reasonable number of organisms grow on each culture plate (dilution plate) or in each tube. The basic assumption is that each propagule in the environment gives rise to one macroscopic colony which can be counted. The assumption is obviously open to criticism; the cells can stay in groups or microcolonies and the medium used will not be suitable for the growth of all micro-organisms. Both these points give an underestimate, and frequently cultural counts are as little as 1 to 10% of direct counts in soil, and even less in water. This may be partly caused by counting dead as well as live cells in the direct method (Table 1.2). The other problem with culturing methods is that with fungi, and to some extent algae, the 'number' really has little meaning. There is no easy way of knowing whether the macroscopic fungal colony arose from a hypha or a spore, and if the former then how large the piece was (i.e. what its biomass was). Careful studies of the origin of fungal colonies have shown that most arise from spores which are dormant in the environment. Repeated counting by dilution plates, over a long period of time such as a year, of dormant microbial cells may give a reasonable estimate of the

Table 1.2 A comparison of the numbers of bacteria recorded by direct observation and by culturing on dilution plates.

Source of sample	Ratio of direct counts to counts from dilution plates
Bacteria in soil and on roots (1):	
(a) On root surface, plant 17 weeks old	2.6
(b) Soil around roots, plant 17 weeks old	3.1
(c) Uncropped soil, 17 weeks after start of experiment	9.2
Bacteria on root surface, plant 12 weeks old (2)	10.2
Bacteria in marine and inshore waters (3):	
(a) Direct count on material trapped on a Millipore filter	147.0
(b) Direct count on a concentrated water sample	2100.0

(1) Louw, H.A. and Webley, D.M. 1959. *Journal of Applied Bacteriology* **22**, 216–226.
(2) Rovira, A.D., Newman, E.I., Bowen, H.J. and Campbell, R. 1974. *Soil Biology and Biochemistry*, **6**, 211–216.
(3) Jannasch, H.W. and Jones, G.E. 1959. *Limnology and Oceanography* **4**, 128–139.

numbers and activity, for the spores have been produced by living mycelium and will over this time span be a reflection of past and an estimate of future activity. The estimate is not good however for an instantaneous measure of activity. The problem of the sort of medium used and the environmental conditions provided, is a major one in all cultural studies. By definition obligate parasites will not grow. Various incubation temperatures, both aerobic and anaerobic conditions, and different nutrient media have to be used in order to allow the growth of as many different organisms as possible. These problems of medium and the environmental conditions have been taken advantage of in various selective culture methods designed to isolate particular groups of organisms. Thus for algae there is usually no carbon source in the medium and the plates are incubated in the light; nitrogen fixing organisms (Chapter 4) may be isolated on nitrogen free medium where most heterotrophs will not grow. It is also possible to select for some organisms by the use of antibiotics, by the aeration and temperature conditions employed or by pasteurizing the inoculum so that only spores survive.

All these methods will allow counts to be made, but there are also culture methods which set out to give only a rough guide to numbers or even just 'present' or 'absent': these are the enrichment methods. They are used when the micro-organisms under study are rare in the environment so that under any conditions that can be devised they will be lost amongst the many colonies of other organisms. The natural soil or water sample is enriched with a substrate that will favour the required organisms and when they have increased in numbers they can be isolated. For example sulphur oxidizing bacteria (Chapter 5) can be encouraged by the addition of sulphur and calcium phosphate to the soil, or sulphur reducers by incubating water or sediment anaerobically in the presence of high levels of sulphate. The great advantage of cultural methods is that in contrast to direct examination it is usually possible, though difficult, to identify the organisms from the culture obtained. Cultural methods are widely used in microbial ecology, particularly for bacteria, and they are best for comparative studies where it is the relative numbers of similar species that are important rather than a complete enumeration of the population.

The third group of methods attempts to assay the metabolic activity of micro-organisms in the environment. If the size of the population is known then the turnover rate or generation time (the time between successive cell divisions) may be estimated by measuring the rate at which tritiated thymidine is incorporated into the DNA of the population. Similarly the rate of consumption of various substrates may be measured by using radioactive tracers, but the amount of the compound under study and the route by which it is broken down in the environment must be known. Attempts have been made to assay the activity of particular common enzymes, such as dehydrogenases; the assumption is that all actively metabolizing cells possess them and use them to an equal degree. The results are confused by the complexity of natural systems with many possible substrates and other, conflicting, biochemical reactions taking place at the same time. The amount of ATP (adenosine triphosphate) in environments has also been used to assess microbial activity (Fig. 1.1); it is assumed that ATP only occurs in live organisms and that the ATP content of organisms is approximately the same or is known for the particular ones under consideration. The level of

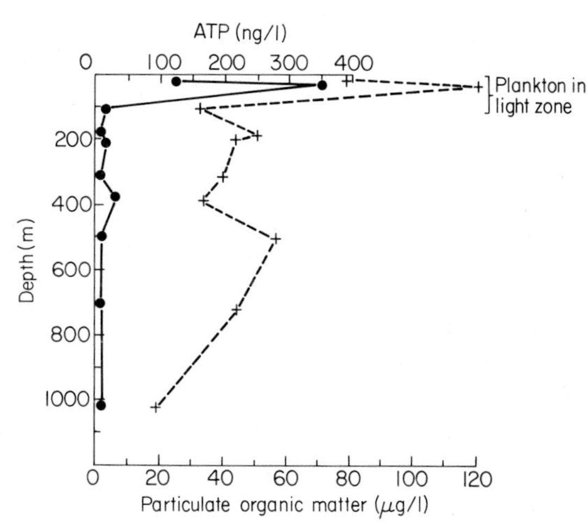

Figure 1.1. ATP (adenosine triphosphate) levels (nanograms/l) at various depths in the Pacific Ocean off the coast of California. The levels of particulate organic matter, which includes the phytoplankton and bacteria, are also shown for the same sampling site. ·———· ATP levels; +– – – –+ Particulate organic matter. (From O. Holm-Hansen, and C.R. Booth, 1966. *Limnology and Oceanography* **11**, 510–519.

ATP is difficult to measure under natural conditions but in the laboratory it is a very sensitive assay (about 10^{-5} μg ATP/l can be detected). The biomass and activity of algae may be estimated by measuring the amount of chlorophyll and other pigments (e.g. Fig. 7.9) and this is usually done spectrophotometrically after extraction with organic solvents. The most widely used method of measuring activity has been to estimate the total respiration, and photosynthesis for algae, as changes in oxygen or carbon dioxide levels. This has met with some success though there are problems such as the occurrence of anaerobic microhabitats, or the use by some micro-organisms of compounds other than oxygen e.g. nitrate,

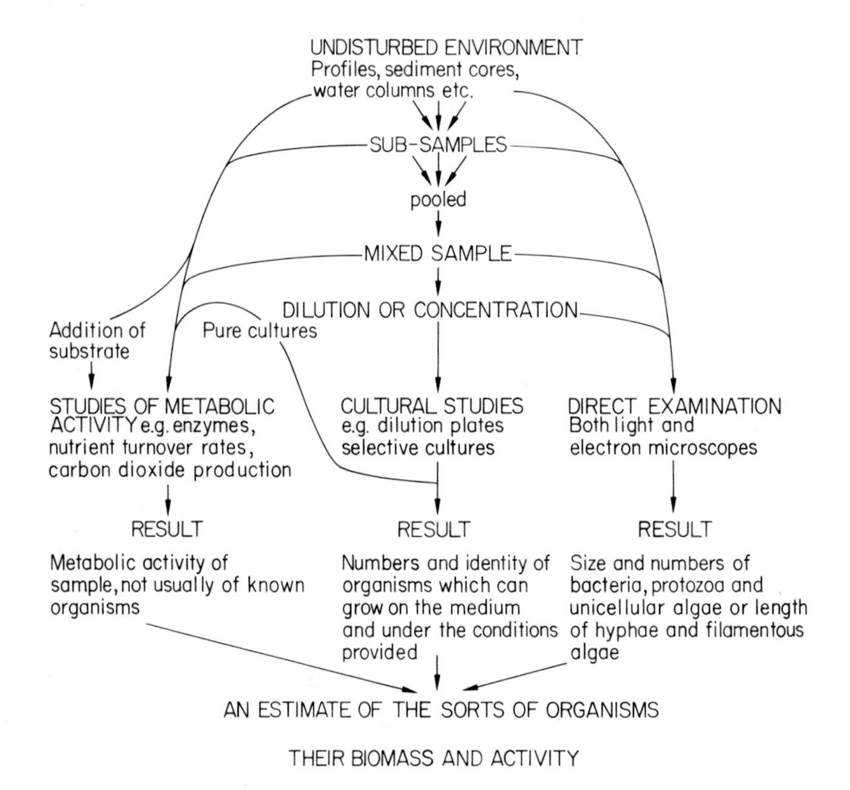

Figure 1.2. Summary of the possible methods of studying the ecology of micro-organisms.

as electron acceptors. All these methods of assaying metabolic activity usually yield no information on the species present in the environment.

This brief review of the general methods (Fig. 1.2) may sound rather depressing. However it is possible to get useful information by combining several different methods, provided that their limitations are realized. Careful consideration of the precise questions to be answered about a particular environment, and the constraints which the environment imposes, can usually lead to the selection of acceptable methods. If microbial ecologists waited for the perfect technique nothing would have been learnt; the incentive to produce better methods is the need to measure particular activities and to solve particular problems.

MODELS IN MICROBIAL ECOLOGY

A rather different approach to the study of microbial ecology has been to set up artificial, usually simplified, environments in the laboratory (page 16). For example the conversion of ammonium to nitrate (Chapter 4) is an important process involving several groups of micro-organisms and it has been studied in soil and water which have been sterilized and then inoculated with known populations of micro-organisms. The rate of the reaction under different conditions can be studied by sampling the system and doing chemical analyses. Similarly by the use of chemostats it is possible to study the way defined mixtures of microbes interact with each other within the controlled environment (further reading 3).

The final 'simplification' is the study of microbial ecology by constructing mathematical models of environments, usually using a computer (further reading 4). The complex natural systems are broken down into many small, more simple, sub-systems and the interactions between and within these are defined in terms of mathematical equations. Usually the models are dynamic so that changes can be studied in relation to time as the various controlling parameters are altered. Basic data are required on the levels of each factor incorporated and its rate of change under the conditions to be considered.

These models do several things for the microbial ecologist: first, and perhaps most important, they force a consideration in detailed, mathematical terms, of just what factors might be important in controlling populations and how these might be linked. Secondly models demand basic data and they usually show the considerable deficiencies which exist in the subject under study. The useful output from models never improves upon the quality of data in the input. Thirdly they generate hypotheses or predictions which can then be tested by experiment: they can therefore point the way to useful areas of study. There are many models for higher plant and animal studies and also for energy and materials flow within habitats but those on the microbial population dynamics in natural habitats are rare and severely limited by the available data.

Having discussed the methods that are available, and their reliability, the next chapter is concerned with how the results that they yield fit in with the general concepts of ecology that have been constructed, mostly with higher plants and animals in mind.

FURTHER READING

1 AARONSON S. (1970) *Experimental Microbial Ecology.* Academic Press, New York. pp. 236.
A comprehensive recipe book.

2 BROCK T.D. (1966) *Principles of Microbial Ecology.* Prentice Hall, New Jersey. pp. 306.
A physiological approach.

3 JANNASCH H.W. & MATELES R.I. (1974) Experimental bacterial ecology studied in continuous culture. *Advances in Microbial Physiology* **11**, 165–212.
Simulations in culture.

4 MAY R.M. (Ed.) (1976) *Theoretical Ecology: Principles and Applications.* Blackwell, Oxford, pp. 350.
A good treatment of computer modelling.

5 PHILLIPSON J. (Ed.) (1971) *Methods of Study in Quantitative Soil Ecology: Population, Production and Energy Flow.* IBP Handbook 18. Blackwell, Oxford. pp. 297.
Methods described and discussed.

6 ROSSWALL T. (Ed.) (1973) *Modern Methods in the Study of Microbial Ecology*. Bulletin 17 from the Ecological Research Committee (Stockholm) of Swedish Natural Science Research Council. pp. 522.
Report of a conference which discussed and described methods.
7 SOROKIN Y.I. & KADOTA H. (1972) *Techniques for the Assessment of Microbial Production and Decomposition in Fresh Waters*. IBP Handbook 23. Blackwell, Oxford. pp. 112.
Methods described and discussed.
8 WILKINSON J.F. (1975) *Introduction to Microbiology*, Basic Microbiology Vol. 1, 2nd Edition. Blackwell, Oxford. pp. 120.
An introduction.

2 Concepts in Microbial Ecology

ENERGY FLOW, FOOD CHAINS, AND TROPHIC LEVELS

Ecosystems are never static. They may be in a dynamic steady state which is a reflection of complex negative feed-back mechanisms controlling the numbers or activity of the individuals, or alternatively the population may change in a way which reflects changing levels of matter and energy within the environment. Energy, mostly in the form of light or heat, originates in the sun, or to a lesser extent from the earth's energy stores, and flows through an ecosystem. It is partly trapped by the organisms within the system and is eventually exported in the form of heat, or chemical energy in reduced compounds. The energy is first trapped by higher green plants, algae and a few bacteria which use light in photosynthesis to reduce carbon dioxide: such organisms are primary producers and they are usually independent of external supplies of organic materials. There are also a small number of primary producers, the chemosynthetic bacteria, which can use chemical instead of light energy to produce reduced carbon compounds. The rate at which energy is stored is called the primary productivity and it may by expressed in energy units (kcal/m^2/day) or as the equivalent amount of organic matter (kg carbon/m^2/day).

The flow of energy through an ecosystem is thus closely linked with that of matter (Chapter 3), though the biomass of the primary producers is not necessarily closely correlated with the productivity, for different organisms have different efficiencies in trapping energy and the biomass represents the results of past energy fixation, not necessarily the present rate. The primary producers are respiring, so the primary productivity may be expressed either as gross productivity equal to the total rate of production, or net when the respiratory loss is allowed for.

There are two main ways in which the primary production of an ecosystem may be used by heterotrophic organisms. Firstly it may be eaten while still alive, by primary consumers (herbivores) who may themselves be eaten by secondary consumers (carnivores): this sequence is a grazing food chain (Fig. 2.1) or a grazing food web if the relationships are so complex that it is not a linear sequence. Secondly if the primary producers are not eaten alive then their bodies, and those of the primary and secondary consumers, enter the detritus food chain (Fig. 2.1).

In aquatic habitats phytoplankton (microscopic suspended or floating algae and blue green algae) are often the major primary producers, though in shallow waters the bottom-growing (benthic) macrophytes, such as the large seaweeds or freshwater plants, may be more important. The main primary consumers in water are zooplankton and in the sea most of the primary production (90%) goes into these consumers (Table 2.1) and the rest of the grazing food chain.

Table 2.1 The different routes by which gross primary production may be distributed in different ecosystems. The gross primary production is taken as 100% and the rest partitioned accordingly: notice the varying importance of the grazing and detritus food chains. Respiration figures are usually based on carbon dioxide production in the dark, and gross primary production figures do not take account of photorespiration (for further consideration of this aspect see I. Zelitch (1971) *Photosynthesis, Photorespiration and Plant Productivity*. Academic Press, New York and London).

	Marine macrophytes in inshore waters Ref. 1 %	Marine plankton in ocean Ref. 2 %	Fresh water plankton (Loch Leven) Ref. 3 %	Fresh water plankton (Lake George) Ref. 3 %	Fresh water plankton and macrophytes (Cedar Bog Lake) Ref. 4 %	Grazed meadow Ref. 2 %	Beech Wood Ref. 2 %
Gross primary production	100	100	100	100	100	100	100
Respiration by primary producers	14.2	8.5	40.3	81.1	21.0	10.3	42.6
Consumption of primary producers by herbivores: this passes along a grazing food chain	11.2	90.3	11.9	17.4	13.3	33.3	40.4
Organic matter released by the decomposition of dead primary producers and dissolved organic matter from live primary producers	74.6	1.2	47.8	1.5	65.7	56.4	17.0
Comments	Half of the organic matter release is from live primary producers.	Grazing food chain is most important.	Detritus food chain is most important.	Very high respiration under tropical conditions	Of the 65.7% organic matter 63.2% is insoluble undecomposed sediment, giving an accumulation of organic matter	More than half of primary production goes directly into the detritus chain.	Much of the primary production is lost in the respiration of the large standing biomass.

1. Khailov, K.M. and Burlakova, Z.P. 1969. Limnology and Oceanography. **14**, 521–527.
2. Gray, T.R.G. and Williams, S.T. 1971. 21st Symposium Soc. General Microbiology. Microbes & biological productivity. Cambridge Univ. Press. 255–286.
3. Open University. 1974. Ecology; Energy Flow through Ecosystems. Unit 5; Whole Ecosystems. Open Univ. Press, Milton Keynes, Eng.
4. Linderman, R.L., 1942. Ecology **23**, 399–418.

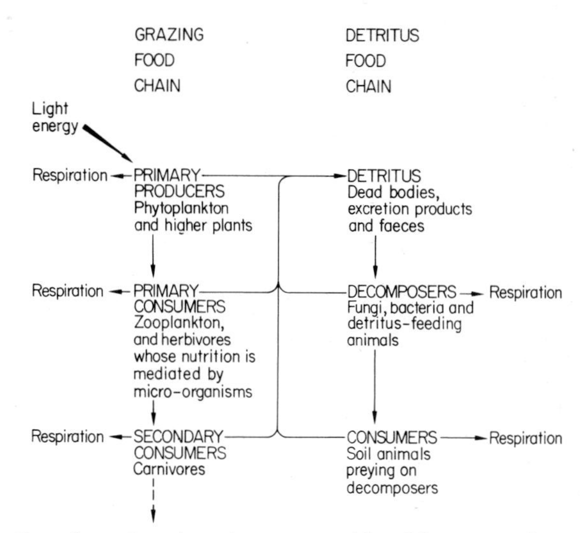

Figure 2.1 The grazing and detritus food chains. See Figure 2.2 for an expansion to a detritus food web.

In fresh waters, and some studies of marine situations, a considerable proportion of the primary production goes into the detritus chain either as particulate material or as exudates (Table 2.1). In terrestrial habitats micro-organisms are not important primary producers nor are they generally considered important as consumers in terrestrial grazing food chains. Much less of the primary production on land goes to herbivores (Table 2.1) but it is interesting that most herbivorous animals depend on micro-organisms in their rumen or caecum to decompose the plant material into a usable form. The grazing food chain, even on land, could thus be dependent on micro-organisms. Most of the energy in a grazing food chain is in the bodies of the organisms involved.

In the detritus food chain, where decomposition occurs, the micro-organisms may not be the dominant organisms; most of the breakdown may be done by arthropods and other small animals (Fig. 2.2), which also play an important part

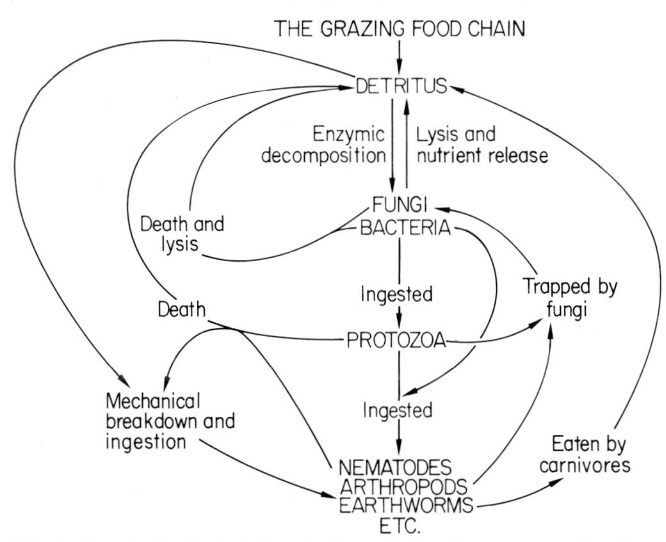

Figure 2.2. The detritus food web depicting the interrelations between organisms by which materials are cycled in the environment.

by comminuting organic matter, exposing new surfaces for microbial colonization. Their faeces are excellent substrates for micro-organisms, having been enriched with various organic compounds and water during their passage through the gut, and microbial activity as assessed by carbon dioxide evolution may be one or two orders of magnitude greater on faeces than on the original detritus. Usually in the detritus food chain bacteria and fungi are responsible for 90% of the energy flow. Most of the energy in a detritus food chain is stored in the environment rather than in the organisms themselves. The chain is dependent on primary production outside itself, generally in a grazing food chain, which may be in another habitat, or be separated from the detritus chain in time when previously produced organic matter like peat is being used. The detritus chain is not solely concerned with decomposition, it contains predators which live, at least partly, off decomposer organisms (Fig. 2.2) and there are also links with carnivores in grazing food chains; earthworms or beetles are eaten by foxes which also prey on herbivores like rabbits. Food chains and food webs are concepts created by ecologists; in the actual environment organisms may not be classified so easily!

There are obviously basic differences in the partitioning of primary production in marine, fresh water and terrestrial environments, and also variation within these environments depending for example on water depth and whether they are in tropical or temperate regions.

The primary producers and the various sorts of heterotrophs in a food chain occupy different trophic levels. In the grazing chain micro-organisms may occupy the lower levels, as primary producers and consumers, while animals occupy the higher levels, but micro-organisms also decompose the bodies from all levels in the chain. The higher the trophic level the less energy it receives, because of respiratory losses, and therefore the biomass is usually, though not invariably, less than that of the lower levels (Fig. 2.3A). The proportion of the primary production that enters the different levels varies with the environment and not all the environments have all the possible levels. For example an environment in which only the detritus chain is operating has the primary production elsewhere, or there may be only primary producers and decomposers with no consumers.

The pyramids of trophic levels based on biomass (Fig. 2.3A) are not however the whole answer. As stated previously (Chapter 1) the metabolic activity in the ecosystem may be what really concerns us: an energy pyramid (Fig. 2.3B) for the same ecosystem as shown in Figure 2.3A illustrates this point, particularly for the decomposers. Thus the decomposer biomass is only about 0.6% of the primary producer biomass per square metre, but some 25% of the energy flow originating in the primary producers passes through the decomposers in this ecosystem.

Though micro-organisms are particularly concerned in these systems as primary producers or decomposers (saprophytes) they can also be pathogens: they can obtain their nutrients direct from live primary producers, herbivores or carnivores. There are varying levels of pathogenicity and of host resistance so there is a complete spectrum from decomposer saprophytes to obligate parasites and even a single species may be a saprophyte in one situation or on one host, but a parasite in or on another (e.g. many fungi). The trophic levels are not therefore self-contained. This movement between trophic levels may involve

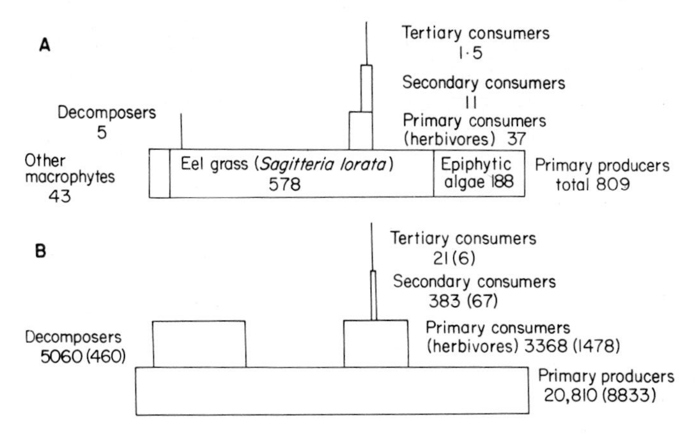

Figure 2.3A. Biomass pyramid for Silver Springs, Florida. The figures are grams dry weight/m^2. The decomposers have a small biomass but high metabolic activity; compare with Figure 2.3B. The horizontal scale is approximately 160 gm/m^2 to 1 cm. (From Odum, H.T. 1957. *Ecological Monographs* **27** (1) 55–112).
B. Energy pyramid for Silver Springs, Florida. The figures are the gross primary production or consumption, in kcal/m^2/yr. The portion of the total energy flow through the group of organisms which is actually fixed as organic biomass and which is potentially available as food for other populations at the next trophic level is indicated in brackets. The horizontal scale is approximately 3800 kcal/m^2/year to 1 cm. Note the importance of the decomposers when their activity is expressed in energy terms rather than biomass as in Figure 2.3A. (From Odum, E.P. 1959. *Fundamentals of Ecology*. W.B. Saunders Co., Philadelphia and London).

quite fundamental changes in metabolism; for example some algae, which are normally photosynthetic primary producers, can become heterotrophs when put in the dark with organic matter and they then become part of the decomposer system. There are even fungi that supplement their nitrogen supplies by catching protozoa and nematodes in special traps (Fig. 2.2), and this presumably puts these fungi, when they are not being decomposers, at the same trophic level as carnivorous animals.

NUTRIENT CYCLES

Though energy flows in one direction through an environment the nutrients are often recycled repeatedly. The movement of many substances can be considered in this way, as biogeochemical cycles (Fig. 2.4), and may involve plants, animals, micro-organisms and physical and chemical reactions. A cycle can be described in terms of pools of the substance concerned, so that there is perhaps a pool of soluble material, a pool of organically bound material and an abiotic, atmospheric pool of the gaseous form (Fig. 2.4 and for example the nitrogen cycle, Chapter 4). The sizes of the various pools may be measured on a local, or even on a biosphere scale. However it is the flow rates between pools which are a measure of the rate at which material is recycled. Even though there is a small pool size, that form of the substance may be very important in the environment: it can be removed from the pool as fast as it enters and the most important forms of a substance, with the highest flow rates, may well have the smallest, and apparently least important pool size. Another way of looking at pool sizes and flow rates is to measure the turnover rates of a pool, that is the time it takes for a unit of material to pass through the pool. The rate at which the whole cycle operates is determined by the slowest turnover rate: there are bottlenecks in the flow within cycles such as the decay of cellulose in the carbon cycle (Chapter 3) or the fixation of atmospheric nitrogen (Chapter 4) in a nitrogen cycle without man-made fertilizers.

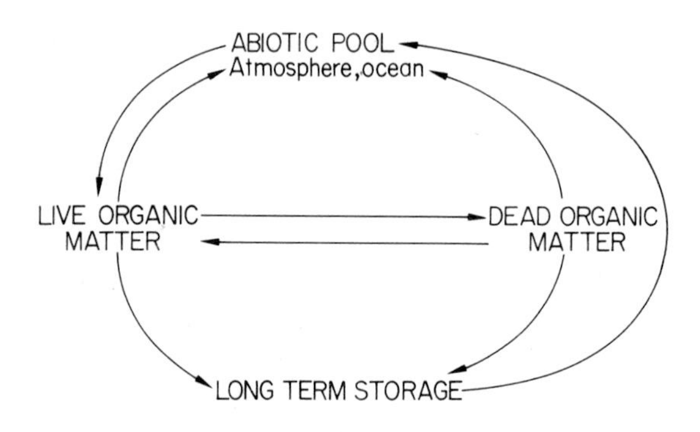

Figure 2.4. A generalized biogeochemical cycle.

There is a tendency, for clarity and ease of description, to consider the cycles in isolation so that one talks about the carbon cycle, the nitrogen cycle, the water cycle and so on. These are all interlocked and interdependent within the environment; the three mentioned would meet in the common organic matter pool where carbon, nitrogen and water are combined in protoplasm. Similarly the decay of organic matter involves the simultaneous mineralization of carbon, nitrogen, phosphorus, sulphur, etc. The cycles are inseparable in nature and the flow rates and the pool sizes of one cycle can influence those of another.

The biogeochemical cycles usually include negative feedback control mechanisms so that any abnormally high pool level or flow rate corrects itself and the system returns to the equilibrium state. Over a long period of time, even as long as the geological time scale, the various interlocked cycles of matter and energy must be in equilibrium. If this was not the case then the cycles would eventually stop. One of the concerns at the present time with 'ecology' is that we do not know enough about the normal pool sizes and turnover times of the biogeochemical cycles: man is now moving about, destroying or dumping large amounts of material which on a local scale are enough to affect the cycles. Whether or not we are exceeding, or have exceeded, the compensating capacity of the global cycles is unknown; the compensating power of a cycle cannot be exceeded without serious and possibly irreversible changes to the environment. What is called in general 'pollution' is either excessive energy flow or the distortion of nutrient cycles by excessive quantities or unusual types of material.

The maintenance of life therefore depends on the continued recycling of inorganic materials and the decomposition of organic matter to produce substrates which other organisms require. The growth of an organism, or the size and productivity of a community, will be limited by the environmental factors, either biotic or abiotic, which are nearest to the critical minimum level required to support life. Thus low levels of oxygen and available nutrients frequently limit the growth of aerobic organisms. Organisms are affected by high as well as low levels: potentially toxic substances may reach damaging levels or dissolved materials in salt lakes may cause such high osmotic pressures that water becomes physiologically unavailable.

INTERACTIONS BETWEEN MICRO-ORGANISMS

Provided that environmental conditions are not so extreme that only a few specialized organisms can survive, there will be competition amongst the members of a community for those nutrients and energy sources which are in short supply. If the abiotic factors and the nutrient supply are favourable then micro-organisms will multiply and compete for space. The total amount of space available is not usually limiting but there are a limited number of suitable micro-habitats which are supplied with nutrients etc.; in such situations there may be competition for usable rather than total space. There are many characteristics of a micro-organism which make it a good competitor including (i) undemanding nutrient requirements and a wide tolerance of environmental conditions so that it can continue to grow and reproduce on a wide variety of substrates and under a variety of conditions; (ii) rapid growth rates and rapid reproduction so that it quickly occupies any available microhabitat, has dormant stages present in almost all places and disseminates itself widely i.e. it has a high inoculum potential; (iii) the ability to produce toxins e.g. antibiotics, to exclude competitors from the microhabitat it is colonizing and at the same time a tolerance of the toxins of other organisms so that it may colonize occupied microhabitats. There are also organisms that develop extreme tolerance to environmental conditions so that they can avoid competition by living in habitats not used by other organisms (such as algae growing on snow or Cyanophyta in salt lakes or thermal springs) or use substrates not degradable by the majority of the population.

There are many other sorts of interactions apart from competition. The general term for all these is symbiosis, though this is sometimes used as though it was synonymous with mutualism in which both organisms are stimulated. In commensal interactions one organism is stimulated while the other is not apparently affected. In parasitism and predation one of the organisms gains benefit and the other is harmed. There can also be no interaction at all, and the relationship is then called neutralism. This covers the spectrum of possible reactions, though finer subdivisions have been devised (and given yet more names!). It is important to stress that it is a continuous spectrum of relationships.

As environmental conditions change, different micro-organisms will be best fitted to exist in the new environment. There are therefore successions of organisms over both short and long periods of time. A succession is the replacement of one community by another as the conditions within the habitat change. The changes may be brought about by the organisms themselves, such as reduction in nutrient and oxygen levels or changes in pH, or they may be imposed from outside the habitat in the case of climatic factors etc. The changes in populations are rarely sudden and species will change in their relative importance within the community until the changes are great enough for a new type of community to be recognizable. Thus there is a sequence of algae in a lake during the year as the water temperature and the nutrient status change (Chapter 7): the fungi and bacteria colonizing dead leaves will change as first the easily metabolized substances are used and then the cellulose and lignin are hydrolysed by a different group of species (Chapter 6). Long term, seral, changes occur in populations of micro-organisms just as in the macroscopic plants and animals; for example bare rock and sand have a succession of communities, usually starting with algae or lichens, as they are converted to stable soil (Chapter 6). The selective pressure at the start of a sere is likely to be great (e.g. very low nutrients) and the species which can survive are few: as the habitat is colonized, diversified and ameliorated the number of species able to grow increases. The final stage in a sere, the climax, is a dynamic equilibrium between the organisms and their environment, and usually shows great species diversity.

THE RELEVANCE OF LABORATORY STUDIES

Laboratory studies have their place, as simplified systems which there is some hope of understanding. However micro-organisms are normally involved in food webs and interlocking nutrient cycles, often competing in crowded, nutrient deficient and antagonistic environments. These situations are obviously very different from growth in axenic culture on carefully prepared nutrient solutions under controlled conditions of temperature, light and aeration (further reading 3). Generation times of micro-organisms are often much longer in natural environments; laboratory studies give times of minutes or a few hours for bacteria (30 min for *Escherichia coli*, 2 hours for *Leucothrix mucor*) but they may be many hours or even days in the natural environment (12 hours for *E. coli*, 11 hours for *L. mucor*). Furthermore, because of the diversity of microhabitats, there can be much more variation in generation times within a natural community than in the more uniform conditions of culture. Care should therefore be used in applying knowledge of growth rates, nutrient requirements, rates of nutrient turnover or competitive ability which have been obtained in the laboratory to more natural situations.

FURTHER READING

1 BROCK T.D. (1966) *Principles of Microbial Ecology*. Prentice-Hall, New Jersey. pp. 306.
Effects of environmental factors on microbes.
2 CLAPHAM, W.B. (1973) *Natural Ecosystems*. Macmillan, New York. pp. 248.
General ecology.
3 DAWES I.W. & SUTHERLAND I.W. (1976) *Microbial Physiology*. Basic Microbiology Vol. 4. Blackwell, Oxford. pp. 185.
Growth rates and growth dynamics in laboratory culture.
4 ODUM E.P. (1971) *Fundamentals of Ecology*. 3rd Edition. Saunders, Philadelphia & London. pp. 574.
General ecology.
5 PHILIPSON J.O.T. (1966) *Ecological Energetics*. Institute of Biology: Studies in Biology. I. Arnold, London. pp. 57.
Energy flow in ecosystem.
6 STRICKLAND J.D.H. (1965) *Production of organic matter in the primary stages of the marine food chain.* In J.P. Riley and G. Skirrow. Eds. Chemical Oceanography. Vol. I. Academic Press, London and New York. 478–610.
Primary production in oceans.

3 Microbial conversions of carbon in the environment

Carbon is one of the major components of living organisms: approximately 50% of the dry weight of organic matter is carbon. Its cycling in the environment is closely linked with the flow of energy, for the major energy stores of organisms are reduced carbon compounds which have ultimately been derived from either protosynthetic, or much more rarely chemosynthetic, fixation of atmospheric carbon dioxide. The animals and plants will eventually die and be decomposed by micro-organisms, particularly bacteria and fungi, which return the carbon to the environment as carbon dioxide. This is the basic carbon cycle, the details of which are the subject of this chapter.

THE BIOGEOCHEMICAL CYCLE

Let us first consider the carbon flow as a global system to obtain some idea of its scale. Figure 3.1 outlines the main carbon pools in the biosphere and gives estimates of some of their sizes and rates of flow between them. Although different authorities vary in the absolute values of these estimates, there is general agreement on the order of magnitude and on the relative sizes of the pools and flow rates, except perhaps for the coal and oil pool where intensive exploration is increasing the estimated size, and figures up to 3×10^{15} metric tonnes have been quoted. More recent estimates also put the total oceanic productivity as less than the total terrestrial, mostly because of an increase in the recent estimates for the latter (e.g. up to 115×10^{9} tonnes/year, see Table 3.1).

The most obvious feature of the cycle (Fig. 3.1) is that the majority of the carbon is not in circulation but is in sediments and storage products such as carbonates in rocks and reduced organic compounds in coal and oil. The carbonate may be precipitated by various organisms such as corals, arthropods, crustaceans and molluscs with calcareous skeletons or shells. Algae may also precipitate carbonate during photosynthesis (Chapter 7). There is a vast buffering capacity within the system due to carbon dioxide dissolved in standing water, mainly in the oceans. Though there will be local variations in atmospheric carbon dioxide in response to changing populations and levels of activity of photosynthesizing and respiring organisms, the overall level of atmospheric carbon dioxide is kept almost constant (though see page 32).

There is great variation in the productivity between and within terrestrial and aquatic environments. This depends on the land use, the depth of the water, the nutrient status and the geographic position; these are summarized in Table 3.1.

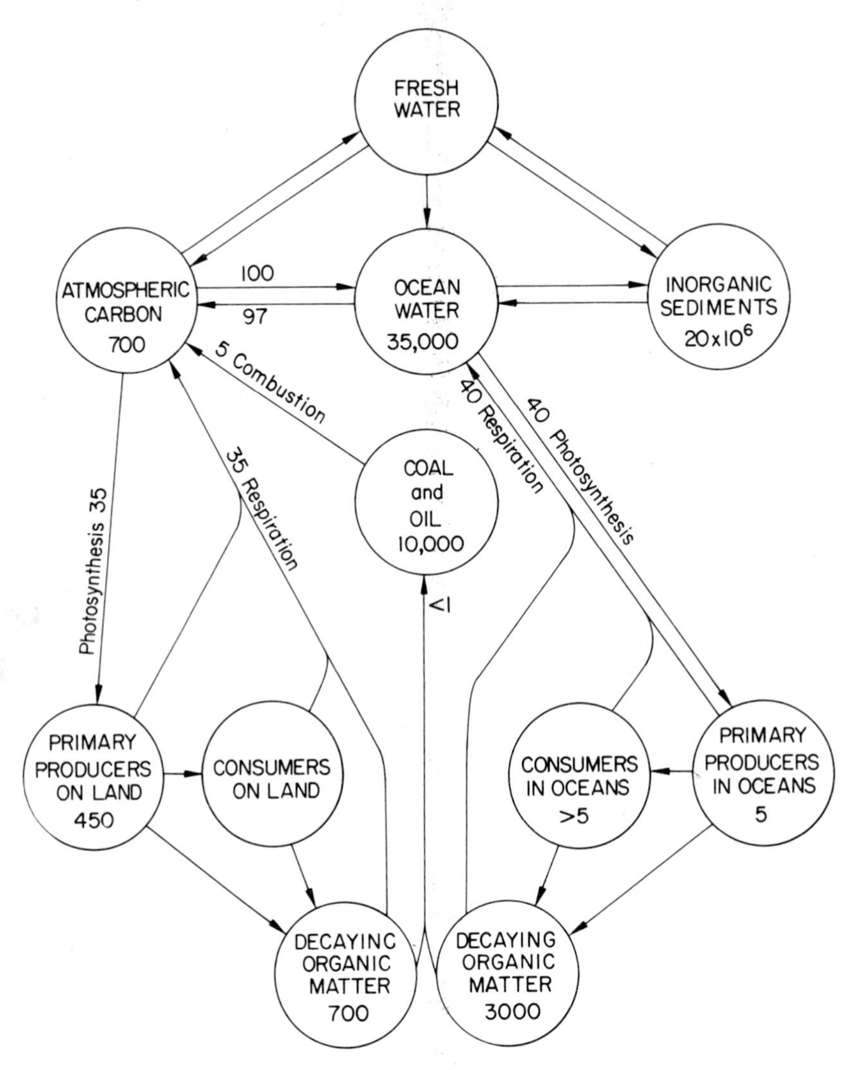

Figure 3.1. Diagram of the carbon cycle on a global scale. The figures given are the estimated sizes of the pools ($\times 10^9$ metric tonnes) or the estimated flow rates between pools ($\times 10^9$ metric tonnes/year). (Data from B. Bolin (1970). *Scientific American* **223**, 125–132.)

THE CARBON CYCLE ON A LOCAL SCALE

The system described above has been built up from the study of many small, relatively defined, ecosystems such as particular ponds, lakes or river systems, or particular agricultural or natural vegetation types. Microbiologists are concerned either with microbial decomposers or with the algal primary producers; these two aspects will now be considered in detail for soil (Fig. 3.2) and water.

The carbon cycle in soil

The primary producers (Fig. 3.2) in soil are mainly higher plants, though the

Table 3.1 World distribution of mean net primary production. Much higher figures may be obtained over short periods, for example in summer even the tundra regions may be quite productive but for a considerable period of the year their productivity is essentially zero. (From R.H. Whittaker (1975) *Communities and Ecosystems*. 2nd Edition. Macmillan, New York. pp. 385.)

Ecosystem	Mean dry weight production g/m^2/yr	World net dry weight production 10^9t/year
Tropical rain forest	2200	37.4
Tropical seasonal forest	1600	12.0
Temperate evergreen forest	1300	6.5
Temperate deciduous forest	1200	8.4
Boreal forest	800	9.6
Woodland and shrubland	700	6.0
Savanna	900	13.5
Temperate grassland	600	5.4
Tundra and alpine	140	1.1
Desert and semi-desert scrub	90	1.6
Extreme desert (rock, sand, ice)	3	0.07
Cultivated land	650	9.1
Swamp and marsh	2000	4.0
Lake and stream	250	0.5
Total continental	**773**	**115**
Open ocean	125	41.5
Upwelling zones	500	0.2
Continental shelf	360	9.6
Algal beds and reefs	2500	1.6
Estuaries	1500	2.1
Total marine	**152**	**55**

Cyanophyta, and to a lesser extent the eukaryotic algae, may be important in a few situations such as rice paddies, on eroded soils and during soil formation (e.g. *Chlorococcum, Palmogloea, Nostoc, Gloeocapsa, Anabaena*). Photosynthetic bacteria, such as *Thiocapsa* and *Rhodopseudomonas*, contribute very little to the total carbon fixation. The consumers in soils are mostly metazoa; protozoa are probably not very important.

The carbon dioxide in the soil atmosphere is usually at greater concentration than in the air, but there is exchange between the two pools by both diffusion through the soil air spaces and also via the soil water and bicarbonate equilibria.

Organic matter reaches the soil from the higher plant primary producers either from their dead leaves, roots and stems (known as litter) or from root exudates. The amounts of litter produced and the mechanism of decay have been extensively studied for many different habitats (further reading 4). The addition of litter may be direct in leaf fall and root death or via the bodies and faeces of consumer animals, including grazing mammals and soil arthropods, earthworms etc. The organic matter can be considered as three main pools in the soil, the insoluble, soluble and microbial pools. The insoluble carbon includes the cellulose and lignin of plant cell walls, chitin in the exoskeletons of arthropods and the walls of some fungi, and other substances which require enzymic breakdown before they yield usable metabolites. Humus is part of this pool and in some

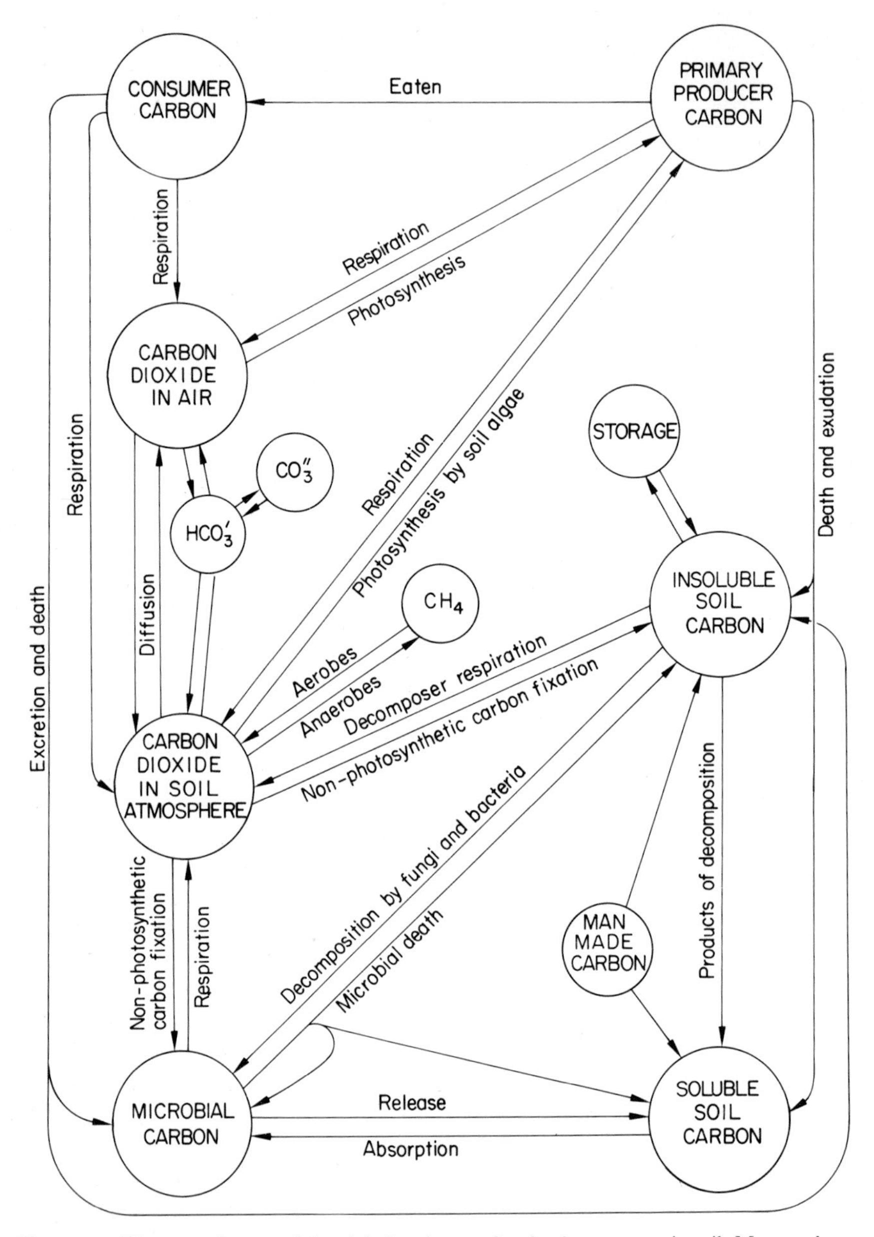

Figure 3.2. Diagram of a part of the global carbon cycle: the decomposers in soil. Man made carbon includes pesticides and other 'artificial' residues.

situations, such as under conditions of acidity or poor aeration, a considerable quantity of insoluble carbon may enter storage as peat. Soluble carbon is that which is in a form immediately available to other organisms and it is either released by live organisms e.g. plant roots (Chapter 6, pages 78–87) or after the bodies of the primary producers, consumers and decomposer microbes themselves are broken down. A considerable portion of the soluble carbon is temporarily immobilized in the cells of the decomposer micro-organisms and its concentration is usually low. This does not mean that it is not important; it is not the concentration of a compound that is critical, but the rate of supply (page 13). The microbial pool is itself small, but rapidly turning over when the micro-organisms are active, and almost all organic matter passes through the microbes during its decomposition before being redistributed to the other pools.

There are many minor reactions within the cycle such as the production of methane (Fig. 3.2), by *Methanobacillus* for example, and its later oxidation under aerobic conditions by *Methanomonas*. Some of the possible reactions are summarized by the equations:

$$CO_2 + 4H_2 \rightarrow CH_4 + 2H_2O$$

$$\underset{\text{(propionic acid)}}{4C_2H_5COOH} + 2H_2O \rightarrow \underset{\text{(acetic acid)}}{4CH_3COOH} + CO_2 + 3CH_4$$

$$CH_3COOH \rightarrow CH_4 + CO_2$$

$$2CH_4 + O_2 \rightarrow \text{intermediates include dimethyl ether, methoxymethanol and methyl formate} \rightarrow 2CH_3OH$$

Other interconnections could be added to Figure 3.2, such as the production of carbon dioxide by the burning of peat or timber from primary producers.

The carbon cycle in water

The carbon cycle in aquatic environments shows a similar set of inter-relations to that just described for soil, though there are differences in emphasis. Firstly there may be a great spatial separation of components; this includes such things as the movement of organisms and material in rivers and the sedimentation in deep waters so that some of the decomposers may be in the bottom muds while the primary producers are many metres above in the upper water layers. Secondly, though it is possible to distinguish several different layers within a large lake or the currents moving through the oceans, aquatic environments (Chapter 7) tend to be more homogeneous than soils where there are many small microhabitats with varying nutrient status (Chapters 1 and 6).

The complete absence of gaseous forms of carbon dioxide and oxygen in water has an effect on the carbon cycle and on the ecology of aquatic environments in general. The 'carbon dioxide in the soil atmosphere' of Figure 3.2 is replaced by dissolved bicarbonate which assumes a central role in the exchange of carbon within the habitat and between it and the atmosphere (Chapter 7). Even though it is quite soluble, the rate of diffusion of carbon dioxide is 10,000 times slower in water than in air, and especially in nutrient rich water it is possible that the availability of carbon dioxide or bicarbonate may limit productivity. Most deep waters have a very low oxygen tension if they are not actually anaerobic, and the decomposition of organic carbon is much slower than under aerobic conditions and will follow different routes. The variation in oxygen and carbon dioxide levels with depth depends on many other factors which will be discussed in Chapter 7.

The primary producers in aquatic systems are mainly phytoplankton though in shallow waters, both fresh and saline, macrophytes may be important. Chemosynthetic and photosynthetic bacteria, as in the soil, contribute little, though there are a few situations such as shallow, nutrient-rich, anaerobic pools where they may produce a considerable fraction of the fixed carbon. The products of photosynthesis pass to the environment in two main ways. Firstly from the exudation of soluble compounds like sugars, amino acids and organic acids (especially glycollic) which usually amounts to 5 to 10% (e.g. 7 μg carbon/m^2/day) of the gross primary production for fresh water plankton and benthos, though higher figures have been reported in very nutrient-poor waters. Marine macrophytes exude up to 37% of their photosynthate. Secondly the primary production may become particulate organic matter either by the death of small algae or by the disintegration of macrophytes, especially when wave action is strong.

This organic matter produced *in situ* (autochthonous material, further reading 4) differs, in chemical composition and importance, in different waters from the organic matter derived from outside the immediate environment (allochthonous). The natural allochthonous material such as leaves or organic matter washed from soils or peat is generally lower in nitrogen than autochthanous material. In fairly still waters, like lakes, the amount of particulate organic matter which is held in suspension is limited, so that as the organic matter level of the water rises, the suspended matter becomes relatively less important and the dissolved material is the major component. Dissolved organic matter is usually more easily decomposed than the particulate as it is largely low molecular weight carbohydrates, but the protein content, and therefore the nitrogen level, is low. These points are illustrated in Table 3.2.

Table 3.2 Approximate levels of organic constituents in lakes with varying organic matter content: the figures are averages for many lakes within each total organic matter class. (From G.E. Hutchinson (1957) *A Treatise on Limnology*, Vol. I. Wiley, New York and London, pp. 1015. Based on data of Birge and Juday.)

Total carbon content mg/l	Live and dead particulate organic matter mg/l	Soluble organic matter mg/l	Crude protein %	Ether extract %	Carbohydrate %	C:N ratio
1.0–1.9	0.62	3.09	24.3	2.3	73.6	12.2
5.0–5.9	1.27	10.33	19.4	1.3	79.0	15.1
10.0–10.9	1.89	20.48	14.4	0.4	85.2	20.1
15.0–15.9	2.32	31.30	12.9	0.2	86.9	22.4
20.0–25.9	2.22	48.12	9.9	0.2	89.9	29.0

Much less is known about the organic matter in marine environments (further reading 4). The level of both particulate and soluble organic matter is generally much lower in the open sea (rarely more than 1 mg carbon/l at the surface) than in most lakes or coastal and estuarine waters. The concentration of carbon decreases with increasing depth to several hundred metres, presumably as it is respired by heterotrophs and as photosynthesis no longer occurs because of the low light levels, and thereafter the concentration does not change until the sediment is reached. At these low nutrient levels absorption by heterotrophic bacteria would be very slow and the adsorption of nutrients onto particulate matter becomes important in concentrating those available. The proportion of marine particulate matter that is alive varies, but in the open ocean it is less than 1% to as high as 20%; the living component may rise to 50% in coastal waters and up to 100% in algal blooms. Most organic matter in the sea is, however, in the sediments and the ratio of carbon in sediments: particulate suspended: dissolved: living organisms is approximately 1000:100:10:1. Carbon accumulates in the sediments because the low oxygen tensions at great depths result in slow decomposition, but the major part of the organic carbon is decomposed while still in the surface layers, where turnover times may be only a few days. The nutrient status of the marine particulate organic matter is, on the basis of rather little information, thought to be slightly different from that in lakes, being higher in protein and having less carbohydrate.

Equilibrium in the carbon cycle

The sort of carbon cycle illustrated in Figure 3.2 is not necessarily in equilibrium. In general, the smaller the scale considered the greater the fluctuations. Globally, (Fig. 3.1) the pool sizes of carbon are so large that their turnover rates are very slow, being measured in tens, hundreds or even thousands of years. A large lake is subject to seasonal fluctuations in environmental factors and in the amount of carbon entering or leaving the system, but it is really a rather stable environment. A small pond is much more variable and at the level of a small crumb of soil the fluctuation and changes in direction within the cycle can be very rapid.

The direction and rate of flow of the carbon cycle is primarily determined by those environmental factors which affect the numbers and activity of micro-organisms. A few examples will illustrate this (further reading 3). *Light* is necessary for a complete cycle since photosynthesis is the major route of carbon fixation. The rate of chemical reactions, including carbon fixation and decomposition, is directly related to *temperature*: to go to the extreme, in a frozen pond the rate of carbon turnover is virtually zero. *Water* is indispensable to terrestrial systems and at low levels the activity of organisms is much reduced, or may be zero. A desert is not a productive environment and the carbon turnover is very slow and the pool sizes very small. An excess of water is equally damaging for terrestrial environments, for a waterlogged soil is usually a poorly aerated soil. *Anaerobic conditions* greatly slow down the rate of decomposition so that much of the primary production goes into storage as peat or lake sediments and is only very slowly recycled. *Acidity* is often associated with peat formation and it further inhibits decomposition. In aquatic environments pH not only affects organisms directly but also the equilibrium between carbon dioxide, bicarbonate and carbonate ions (Chapter 7) and hence the availability of carbon dioxide to primary producers. The pH also affects the solubility, and therefore the availability, of many nutrients essential for microbial activity. *Minerals* themselves may be rate limiting in environments with very low nutrient levels. Nitrogen is the nutrient most commonly in short supply during the later stages of plant decay (the insoluble carbon of Fig. 3.2) because there is very little of it in plant cell walls.

THE DECOMPOSITION OF SPECIFIC TYPES OF CARBON COMPOUND

Naturally occurring compounds

When an organism dies the amino acids, peptides and low molecular weight carbohydrates, which mostly occur in the protoplasm, are absorbed directly by the decomposers and rapidly recycled. The high molecular weight compounds which form some storage materials (lipids, starch) or the exoskeletons or cell walls (Table 3.3) are broken down by fungal and bacterial extracellular enzymes to small molecules which can be directly absorbed. Protozoa can ingest, by pinocytosis, much larger molecules and also small particles up to at least the size of bacteria, small algae and fungal spores. The decay of organic materials involves

Table 3.3 The main organic components of exoskeletons and cell walls. + indicates presence of a compound; (+) indicates the presence in small quantities or in only some organisms in the group indicated.

	Organic Constituent								
Organism	Cellulose	Lignin	'Hemicellulose'	Proteins, peptides and amino acids	Mucilage coverings	Mucopeptide (murein, or peptidoglycan)	Lipid	Chitin	Other Polysaccharides
Arthropods				+			+	+	
Crustacea				+				+	
Protozoa				+	(+)			(+)	
Algae	+		+	(+)	+			(+)	+
Fungi	(+)		(+)	+	(+)		(+)	+	+
Bacteria			(+)	+	+	+	+		+
Higher plants	+	+	+	(+)			(+)		(+)

elements other than carbon, but these will also be considered in the following discussion. The fact that nutrient cycles are interlocked with each other has been stressed before (page 14).

Proteins in natural materials are less easily utilized than laboratory studies with pure proteins might suggest. In nature proteins are often complexed with polysaccharides or tannins and are more resistant to decay. Fibrous proteins with many cross-links, such as keratin, are very resistant to microbial attack though most actinomycetes (e.g. *Streptomyces*) and some fungi (e.g. *Penicillium*, *Keratinomyces*) can degrade them. Proteins have the great nutritional advantage for micro-organisms that they contain both carbon and nitrogen.

There is surprisingly little information on the breakdown of *lipids* and *starch* in the natural environment but these common constituents of organisms are readily utilized by bacteria and fungi in the laboratory. Again the rate of decay is probably slower in nature where these substances may be closely associated with ones less easy to degrade. Under anaerobic conditions only some bacteria (e.g. *Clostridium*) can cause the decomposition.

Chitin is an important source of carbon in the environment. It is a polymer of N-acetyl-D-glucosamine and like proteins it contains nitrogen. Decomposition is brought about mostly by actinomycetes (e.g. *Streptomyces*) and other bacteria (e.g. *Pseudomonas*, *Bacillus*, *Clostridium*): the relative importance of each depends on aeration since the actinomycetes are obligate aerobes while some of the other bacteria are anaerobic. In very acid environments fungi (e.g. *Mortierella*) may become more important in chitin breakdown as they are less sensitive to low pH than are most bacteria.

The distinctive polymer in bacterial cell walls, a *mucopeptide* (also called peptidoglycan or murein), is composed of N-acetylglucosamine and N-acetylmuramic acid, with peptides linked to all or some of the latter in some species. It is sometimes not a major component of the walls by weight, but considering the wide distribution of bacteria and their high biomass in some environments, its breakdown is important to the carbon balance. Much is known about the enzymes concerned with bacterial lysis under laboratory conditions but very little about the breakdown in natural environments. It is probable that it is mostly autolysis, though *Myxobacterium* and some *Bacillus* species can cause bacterial lysis. There is also a bacterium (*Bdellovibrio bacteriovorus*) which is parasitic on other bacteria, causing lysis of susceptible hosts, and it is apparently widespread in natural habitats.

The decomposition of *cellulose* has been the subject of many investigations, because it is the major constituent of plant cell walls and therefore of the insoluble carbon added to the cycle, and also because it is widely used by man as textiles, paper, and as a component of timber. Cellulose is a polymer of D-glucopyranose (containing mostly $\beta 1.4$ glycosidic linkages) whose molecular weight is usually about one million. It occurs in plants in a semi-crystalline form. Decomposer micro-organisms can produce an enzyme, cellulase, which catalyses the hydrolysis of the polymer, either endwise or in a random manner, to the dimer cellobiose. The term cellulase describes an enzyme complex which acts in two distinct stages: firstly there is loss of the crystalline structure, with no corresponding loss of strength, and then the depolymerization itself occurs. The resultant cellobiose is hydrolysed by the enzyme cellobiase to glucose which is absorbed by the decomposer or enters the soluble carbon pool. The micro-organisms that carry out this breakdown vary with the environment. Under aerobic conditions a wide range of fungi are important (e.g. *Chaetomium, Stachybotrys, Trichoderma, Penicillium*) though bacteria also occur and may be the major organisms in aquatic environments (e.g. *Cytophaga, Bacillus, Pseudomonas*). Some bacteria can decompose cellulose anaerobically and are therefore important in waterlogged soils and in deep water sediments (e.g. *Clostridium*). Even in well aerated environments, there are local anaerobic microhabitats caused by aerobic decomposers using oxygen at a rate in excess of that supplied by diffusion through the water or water films.

The other important carbohydrate constituents of many plant cell walls are a heterogeneous group of compounds, collectively called hemicelluloses, forming the residue in analysis after cellulose and lignin are accounted for. They include various polymers of hexoses, pentoses and sometimes uronic acids (acids derived by the oxidation of the hydroxyl group at carbon atom 6 of sugars). Commonly occurring members of the group are *xylans*, *mannans* and *pectins*. Many hemicelluloses are quite easily decomposed by bacteria and fungi in the laboratory, but once again in nature they are frequently complexed with other substances and this may make their breakdown more difficult. The microbial attack on pectin (mostly polygalacturonic acid) has been much studied because of the importance of this substance in the middle lamellae of plant cell walls. Plant pathogens and saprophytic microbes often attack this layer and cause the disintegration of the plant tissue. The process is quite complicated and involves several different enzymes collectively known as pectinase. The breakdown of pectic substances in natural environments may be delayed by the repression of these enzymes by high sugar levels in fresh substrates. This is one of the factors responsible for substrate succession (Chapters 6 and 7).

The other main constituents of higher plant cell walls are *lignins*, but their chemistry is much less well understood. They are thought to be polymers of p-hydroxyphenylpropanes and are characteristically difficult to degrade, either chemically or biologically. There are however some fungi, particularly Basidiomycotina (which includes the common mushrooms and toadstools and the bracket fungi on trees) and some bacteria (e.g. actinomycetes) which are capable of decomposing them.

Cellulose and lignins are the main constituents of timber which is an important carbon product of the biosphere and is also a major raw material used by man. Forest ecosystems are very widespread in the world and can have a high productivity (Table 3.1). Recycling of the carbon is therefore important and is mostly due to fungi, particularly Ascomycotina (soft rots) and Basidimycotina (hard rots). Soft rots are important in wet situations especially in small sizes of timber or where there is abrasion e.g. on pilings and piers used by boats. The hyphae occur within the secondary walls of the vessels or tracheids. In hard rots the basidiomycete hyphae grow within the tracheid lumen and only occasionally branch at right angles and pass through the wall by a bore hole. The colour of hard rots is indicative of the chemistry of the decay; in some the wood goes brown as the cellulose is removed and the lignin left, in others a white coloration develops, usually because of the removal of lignin. It is possible to identify some common decay fungi on the basis of their rot type and so be able to prescribe suitable control measures without the tedious cultural identification or the presence of fruit bodies. While this decomposition is essential to the cycle it can also be very damaging economically if the timber is in use. The destruction of housing timbers by dry rot (*Serpula lacrymans*) in Europe or by *Poria incrassata* in North America must be only too familiar to some readers, who may take what comfort they can from the fact that their houses are part of a biogeochemical cycle!

Paper, cotton, linen and jute textiles are also natural plant fibres, largely composed of cellulose. They are decayed during manufacture, storage and use, mostly by ascomycetes and deuteromycetes (e.g. *Chaetomium* and *Stachybotrys*).

The decomposition of some of the compounds discussed above, together with many more easily degraded ones, is of importance in *food* decay. What is good for man is usually good for heterotrophic bacteria and fungi. The decomposition of food during manufacture and storage is a very complex subject (further reading 2) but in general wet proteinaceous foods such as meat, eggs, fish and milk are decayed by bacteria: dry, sugary or carbohydrate foods such as flour, bread, cakes and most fruits are usually decayed by fungi. A distinction should be made between the decay of food which, though it sometimes produces unpleasant tastes or smells, is not dangerous (e.g. the growth of fungi on bread or cheese), and the presence on or in food of dangerous bacteria such as some *Salmonella* spp. or *Clostridium* spp. which cause food poisoning, often with no easily detectable degradation of the food.

Manufactured materials

Virtually all products synthesized by living organisms can eventually be broken down by micro-organisms and in addition some of the organic products of man's chemical syntheses are degradable (further reading 8). Even some of the substances whose main advantage was once thought to be their stability, like *plastics*, are degradable in some circumstances. Plastics are a heterogeneous group of compounds, including polyalkenes (such as polyethelene (polythene), polybutylene, polypropylene), polystyrenes and polyvinyl chloride (PVC). The polyurea and polyformaldehyde hard plastics are also produced in large quantities and polyurethane is yet another type. There are examples of degradation by both fungi and bacteria of all types but rarely to any significant extent under normal conditions of use. The higher the molecular weight of the polymer the less the likelihood of biodeterioration. The only common example is the growth of mould on PVC shower curtains and even here it is often growth on the fillers, dyes or plasticizers (such as dibutylphthalate), which causes the plastic to go brittle, rather than decay of the plastic itself. Under marine conditions there may be extensive surface growth on plastics but again decay by micro-organisms is not extensive. Plastics have been produced in quantity for only a few years and yet there are already indications that micro-organisms may degrade some of these novel substances and recycle the carbon. The lack of rapid breakdown of most plastics is a problem in refuse disposal and there is now the possibility of producing plastics that are partly broken down chemically, by ultraviolet light for example, to small enough molecules for the micro-organisms to complete the decay.

Rubber is subject to microbial breakdown, particularly natural rubbers rather than the synthetic ones like neoprene. This along with the decay of plastics, is serious in electrical insulation of buried cables and in the sealing rings of underground sewage pipes where the seals can decay long before the concrete pipes themselves need replacing. The organisms concerned are fungi and actinomycetes. Some of the accelerators used in the polymerization of rubber, such as dehydroabietyl ammonium pentachlorophenate, can help to prevent decay because they have biocidal properties, or biocides (page 29) may be added during manufacture.

Paints used to be based on heavy metal pigments like white lead and these were toxic to micro-organisms, and also to man. For the latter reason paints do not now contain much lead, except special anti-corrosion and marine fouling prevention paints and some primers. Modern paints are also thickened with methyl cellulose or similar substances to make them 'non-drip' and many are now water based rather than oil. They are therefore subject to decay by bacteria in the anaerobic conditions of full tins in storage, or by both bacteria and fungi in use. They routinely contain a range of biocides to prevent or reduce this decay. Apart from the decay of the paint substance itself, the bondings, usually synthetic resins, which hold the paint film onto the surface can be broken down and the complete paint film peels off. One of the main functions of primers when wood is being painted, is to form a firm, impervious layer between the paint and the substrate so that nutrients for micro-organisms cannot be washed out of the wood.

Finally in the biodeterioration of carbon compounds *petroleum* must be considered. The amount decayed, in relation to the amount used, is very small but the problem is that corrosion occurs during the growth of organisms in metal fuel tanks. This may be by the supply of hydrogen ions, by cathodic depolarization in which the protective layer of hydrogen gas is removed by organisms such as *Desulfovibrio*, or by the production of oxygen depletion cells under microbial colonies. A fungus, *Cladosporium resinae*, has been particularly studied but deterioration is also caused by many other fungi and bacteria, e.g. *Pseudomonas*. Petrol (gasoline) used for cars usually has too much lead additives for decay to occur and the main problem has been the various sorts of aircraft fuel. Fuel tanks contain water from condensation and this allows growth of the micro-organisms with the fuel, or even fuel vapour, as the sole carbon source, with nitrogen and minerals from dirt and impurities. Apart from the corrosion the mycelium can block fuel lines and gauges. Biocides such as the de-icing fluid ethylene glycol monomethyl ether can be used continuously, but are expensive, and control usually depends on good fuel hygiene and tank cleaning during maintenance. Periodic flushing with biocides is also practised and the tanks are lined with corrosion resistant coatings of butyl rubber or epoxy resins. Fungi and bacteria can, and do, attack lubricant oils, cutting oil emulsions and hydraulic fluids. There is not usually much loss of material but there are serious effects on viscocity, emulsion stability and the blocking of pipes and valves. Crude oil decay has become important in natural disposal of oil spills at sea. The lighter fractions evaporate and the remainder is colonized by micro-organisms (e.g. *Pseudomonas, Achromobacter, Flavobacterium, Candida* etc.) which may degrade 40 to 90% of the oil depending on its chemistry and therefore its source. Alkanes and other saturated compounds are degraded first, then aromatic and heterocyclic chemicals. Surfactants and other substances used to disperse the oil may be toxic to micro-organisms and so delay the eventual breakdown.

Much research has been devoted to the prevention of decay when a carbon product is in use. Decay never occurs if the material is kept dry, and this is the major control measure used for many substances e.g. seasoning timber and dehydrating food. It is sometimes possible to change some other environmental parameter to a point outside the normal range for growth. Temperature may be raised to above the thermal death point and then further contamination prevented as in food bottling and canning. Low temperatures retard the growth but in normal home deep-freezes the lower thermal death point is rarely reached. If all else fails, and a non-degradable material cannot be used, then biocides are added. Several specific examples have already been mentioned but in general they depend on heavy metal ions such as chromium, mercury, copper, zinc and boron or on phenolics. Sometimes these are combined with halogens or with arsenic as in the copper-chrome-arsenate wood preservatives or pentachlorophenate which is widely used.

Uses of microbial decay of carbon

Occasionally man deliberately encourages decomposers, notably in sewage works and in some of the modern treatments for composting household garbage (further reading 7). Both these valuable commodities are rich in organic carbon compounds and the treatments are concerned firstly with breaking the material up to increase the surface area open to attack by the decomposer micro-organisms such as bacteria and protozoa, and secondly with supplying sufficient oxygen, for example in activated sludge tanks, to allow rapid aerobic decomposition. Algae growing in sewage ponds may be an important source of oxygen, and in tropical regions they can supply the full requirement. In addition experimental studies suggest that the algal crop could be harvested and used to produce animal feedstuffs.

There are many cases where man attempts to modify the natural decomposition process in order to yield particular end products for his use. Industrial fermentations may use cheap products of other industries as a carbon source, for example waste fractions of petroleum refining and natural gas may be used for protein production. The making of silage is a modification of the natural decay of vegetation; by creating anaerobic conditions enough lactic and acetic acid are produced by fermentation to prevent complete breakdown of the grass.

Biocides and the carbon cycle

Attempts to control the effects of pesticides, herbicides and fungicides on the carbon cycle are not entirely successful. Although they are intended to be persistent, so that they can stay in their appointed place to act for as long as possible, they not infrequently move into the biosphere in general, and if they move up a food chain they become potentially harmful. They are a very diverse collection of chemicals including amongst many others, plant hormone analogues (phenoxyacetic acids), substituted triazines and ureas, bipyridylium compounds, dithiocarbamates, carboximides, heavy metal salts, phenolic compounds, organophosphates and organochlorine compounds like DDT and BHC (see further reading 5 and 7).

Most of these compounds do not occur naturally; some of them are toxic to micro-organisms and resistant to decomposition. This has caused concern, highlighted by the discovery of the persistence and ubiquitousness of organochlorine compounds (e.g. DDT) and their breakdown products in the environment. The use of such chemicals is now discouraged, or even banned in some countries, but if there are no alternatives a balance must be struck between the immediate value to humans in increasing food production and preventing disease, and the long-term potential harm to the whole environment. Persistent chemicals present the problem that they accumulate in organisms near the top of food chains (Chapter 2) to a much higher level than in the environment generally. This seems particularly true of aquatic systems where small molluscs, crustaceans and some fish, may have levels a thousand times greater than the water. Bacteria and algae, at least in aquatic systems, can also accumulate pesticides against a concentration gradient.

A further problem may arise with the repeated use of even a non-persistent, biodegradable chemical. The first, or limited, application may result in some distortion of the microflora which nevertheless recovers and degrades the chemical. Repeated distortion over periods of years may however cause serious damage and this possibility must be considered in the new systems of agriculture which involve no ploughing but which use herbicides such as paraquat, a bipyridylium compound, to remove weeds. Problems have also been caused by impurities in herbicides but this is not usually serious in normal use of commercial products.

Biocides are now usually designed to be biodegradable (most do not last more than a year) and studies suggest that bacteria and fungi are most important in their breakdown though algae may be active in some situations. Even the now notorious DDT, and aldrin and dieldrin, are not toxic to most soil micro-organisms at the normal concentrations used, indeed their application to the soil may increase the number of bacteria that can be isolated. Often a particular organism may only be able to perform one step in the breakdown and the complete removal from the soil or water depends on a series of organisms. Laboratory studies with pure cultures or a limited number of species in mixture may therefore be misleading. A persistent chemical may be altered but not degraded, and the product may be just as difficult to degrade, and sometimes as toxic, as the original compound: thus DDD is more stable and more toxic to micro-organisms than the DDT from which it was derived by microbial dechlorination. Though a compound may not be degraded when presented as a sole carbon source it is often broken down (co-metabolized) when in the presence of utilizable carbon, which is the situation in most environments (Fig. 3.3).

Some of the chemical groupings which are liable to make novel compounds either persistent or biodegradable are now known (further reading 5, 6 and 7). Chemicals which are usually resistant include those with substituted amino, methoxy, sulphonate and nitro groups, chlorine substitution particularly in the meta position, meta substituted benzene rings in general (rather than ortho or para), ether linkages and branched carbon chains (Table 3.4). Quite small chemical changes can have a great effect on the degradability (Fig. 3.3).

The ability of a compound to be degraded or immobilized in the environment can also be affected by the rates of application and the proprietary formulation (the fillers, dispersing and wetting agents etc. that are added to the active chemical): it is therefore important that the recommended conditions and rates of application are followed. There are still many gaps in our knowledge of the effects of these chemicals and there will undoubtedly be unforseen problems with

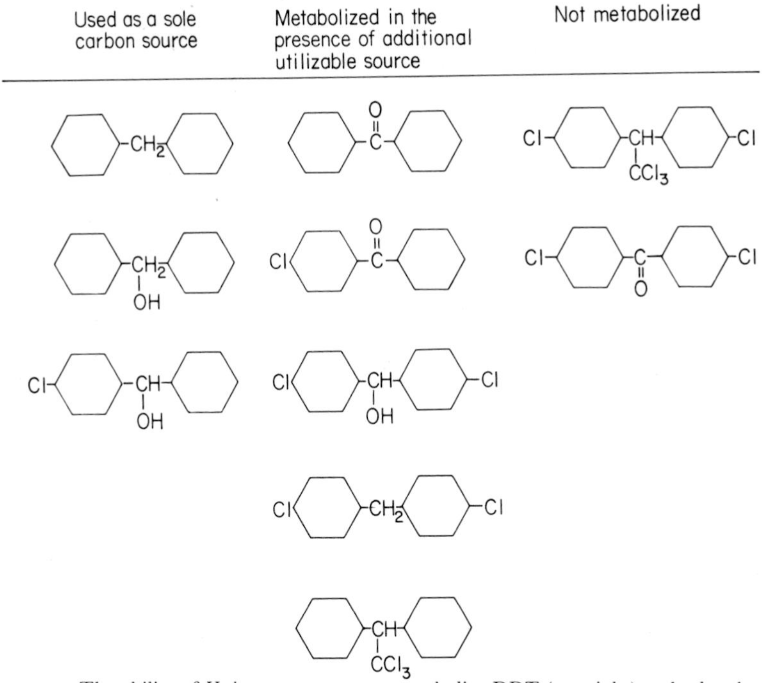

Used as a sole carbon source | Metabolized in the presence of additional utilizable source | Not metabolized

Figure 3.3. The ability of *Hydrogenomonas* sp. to metabolize DDT (top right) and related compounds. (From R.E. Cripps, 1971. Microbial breakdown of pesticides. In G. Sykes and F.A. Skinner, Eds. Microbial Aspects of Pollution. Academic Press. Original data from D.D. Focht and M. Alexander, 1970. *Science, New York* **170**, 91.)

Table 3.4 The decomposition of phenoxyalkyl carboxylic acids in soil. Some of these compounds are used as herbicides. (Modified from M. Alexander and M.J.H. Aleem (1961) *J. Agric. Food Chem.* **9**, 44–47).

Compound	Substitution of chlorine	Days for disappearance from soil	Comment
2-chlorophenoxyacetate	ortho	>205	Anomalous behaviour degradation expected
4-chlorophenoxyacetate	para	11	Carboxylic acid side
3-(4-chlorophenoxy)propionate	para	11	chain has slight effect
4-(4-chlorophenoxy)butyrate	para	53	on rate of decomposition
2,4-dichlorophenoxyacetate	ortho/para	26	Addition of chlorine
3-(2,4-dichlorophenoxy)propionate	ortho/para	4	in ortho position has
4-(2,4-dichlorophenoxy)butyrate	ortho/para	11	little effect
3,4-dichlorophenoxyacetate	meta/para	>205	Addition of chlorine
3-(3,4-dichlorophenoxy)propionate	meta/para	>81	in meta position
4-(3,4-dichlorophenoxy)butyrate	meta/para	>205	increases resistance
2-methyl-4-chlorophenoxyacetate	para	70	Meta chlorine
4-(2-methyl-4-chlorophenoxy)butyrate	ortho/para	39	increases
2,4,5-trichlorophenoxyacetate	ortho/para/meta	>205	resistance

their use in the future, such as those which have already occurred with the organochlorines. In general the amounts used are small on a biosphere scale and most, but not all, are degradable by design or by the fortuitous adaptability of micro-organisms.

THE EFFECT OF MAN ON THE CARBON CYCLE

Toxins in industrial effluent are far greater in quantity, and more concentrated locally, than biocides. They cause distortions of the cycle simply by harming or killing the organisms concerned. However the main effect of human activity on the cycle is by the direct addition or removal of material.

Firstly there is the increase in the carbon dioxide level in the atmosphere, largely caused by the burning of between 5 and 6×10^9 metric tonnes per year of fossil fuels (Fig. 3.1). This has occurred in the last hundred years and particularly in the last few decades. Not all of this carbon dioxide stays in the atmosphere; it equilibrates with the oceans and may also have increased the photosynthetic rate (which is often limited by carbon dioxide) so becoming incorporated into the plant carbon pools. The overall effect on the energy absorbed from the sun by the earth's atmosphere or on the internal equilibria of the planet is unknown; dire predictions of climatic changes, melting ice caps etc. have been made but there are also hopeful statements of increased plant production and a stability of the equilibria.

Other changes caused by man are less noticeable on a global scale, though they may be dramatic locally. For example sewage and major oil spills amount to only a few per cent of the annual carbon fixation by photosynthesis, and are many orders of magnitude less than the present biomass of the standing crops and the organic matter in the sediments. The addition of sewage (page 125) causes a local imbalance of the cycle for the organic matter originated outside the system. In natural conditions the organic matter produced by photosynthesis would be balanced by decomposition, even though each was running at a high level. There is nothing intrinsically wrong with feeding organic matter into the cycle, but it is addition beyond the local capacity that results in pollution.

Modern agricultural practices affect the carbon cycle. There is now a tendency to concentrate animals in intensive units which again leads to severe local problems with organic waste disposal. On mixed farms the animal manure could be spread on the fields but in intensive units this may not be possible, and furthermore the waste is now often a slurry, rather than a solid, and this may adversely affect the soil structure and microbiology (Chapter 6, page 90). The effect of intensive arable cultivation is less easy to understand. A change in the plant species, from natural mixed grassland community to a single species crop, will affect the structure of the soil, the micro-organisms and hence the fertility of the soil and the decomposition of plant remains. This change need not be deleterious as it may reflect, for example, different groups of micro-organisms which are now required to process the different carbon compounds reaching the soil from the new vegetation. Intensive cropping will remove more carbon to distant consumers, whose dung and bodies would previously have returned at least some of the fixed carbon to the soil. This illustrates a general feature:

pollution in one place often means a deficiency in another. The sewage in the rivers implies a deficiency of organic matter and nutrients on the land which we attempt to replace by mineral fertilizers manufactured somewhere else again. Similarly pollution by toxic metals means that they have been removed from some 'safe' store like ore, chemically changed or concentrated and then released where they are not part of the natural cycles.

To summarize the microbiology of the carbon cycle: micro-organisms play major parts as both primary producers in aquatic environments and as the only decomposers in the cycle. This latter role may lead to biodeterioration of materials but it is also directly useful for the removal of increasing quantities of 'waste' produced by modern industry and farming practices.

FURTHER READING

1 Alexander M. (1971) *Microbial Ecology*. Wiley, New York. pp. 511.
General text.

2 Barnell H.R. (1974) *Biology and the Food Industry*. Studies in Biology 45. Arnold, London.
Micro-organisms in food manufature and decay.

3 Brock T.D. (1966) *Principles of Microbial Ecology*. Prentice-Hall, New Jersey. pp. 306.
A physiological approach.

4 Dickinson C.H. & Pugh, G.J.F. Eds. (1974) *Biology of Plant Litter Decomposition*. Academic Press, London. Vol. 1. pp. 146. Vol. 2. pp. 175.
Comprehensive reviews.

5 Higgins, I.J. & Burns R.G. (1975) *The Chemistry and Microbiology of Pollution*. Academic Press, London, New York and San Francisco. pp. 248.
Pesticide and synthetic polymer breakdown, sewage disposal.

6 Laskin A.I. & Lechevalier H. Eds. (1974) *Microbial Ecology*. C.R.C. Press, Cleveland, Ohio, pp. 191.
Presticide breakdown.

7 Sykes C. & Skinner F.A. (1971) *Microbial Aspects of Pollution*. Wiley, New York and London. pp. 416.
Review of pesticides, sewage and waste treatment.

8 Turner J.N. (1967) *The Microbiology of Fabricated Materials*. Churchill, London. pp. 296.
Biodeterioration of materials.

9 Wilkinson J.F. (1972) *Introduction to Microbiology*. Basic Microbiology I. Blackwell, Oxford, pp. 120.
General text.

4 Microbial conversions of nitrogen in the environment

The conversion and cycling of nitrogen in the biosphere is the next most important process after the transformations of carbon. The two cycles are linked by the occurrence of nitrogen in organic molecules: especially proteins. It is therefore an essential, though not necessarily major, constituent of all cells, being from 1 to 10% by weight in plants, and up to 20 or 30% in animals.

THE BIOGEOCHEMICAL CYCLE AND THE DISTRIBUTION OF NITROGEN IN THE BIOSPHERE

The main pools in the environment are as gaseous nitrogen in the atmosphere (about 80% of air by volume) and in geochemical deposits, where it may occur as organic sediments, or as ammonium or occasionally nitrate in various rocks (Fig. 4.1). The amount actually in circulation in the bodies of organisms is a very small proportion of the whole. Most of the flow rates between pools are known less accurately than for the carbon cycle; this is partly a reflection of methodology for it is very much more difficult to work with ^{15}N than with radioactive ^{14}C.

A key point in the cycle is the fixation of gaseous nitrogen into inorganic compounds. Until recently the most significant contribution was biological fixation by various prokaryotes (see below). However industrial fixation, mostly by the Haber-Bosch process, now accounts for about 50% of the nitrogen fixed in the world and though this is mostly applied to the land as fertilizers some of it reaches the sea as river run-off. The Haber-Bosch process fixes nitrogen at 200 to 1000 atmospheres pressure and about 550°C which requires an enormous use of energy: micro-organisms do it at atmospheric pressure and about 5 to 20°C, though it is still a process requiring much metabolic energy (the bond energy of N_2 is 226 kcals/mole). In many ecosystems mineral nitrogen is the most common limiting nutrient for primary producers. The amount of nitrogen in the ocean flora and fauna is low and this is a reflection of the lower biomass per unit volume (Fig. 3.1).

Nitrogen in aquatic environments

In fresh water (further reading 4) the distribution of nitrogen depends on the particular body of water studied, but usually nitrogen gas is dissolved in the surface waters to levels approaching saturation and in deeper water it may be supersaturated. The levels of mineral nitrogen vary seasonally and are largely determined by the run-off from soil, and the amount added in rainwater. There

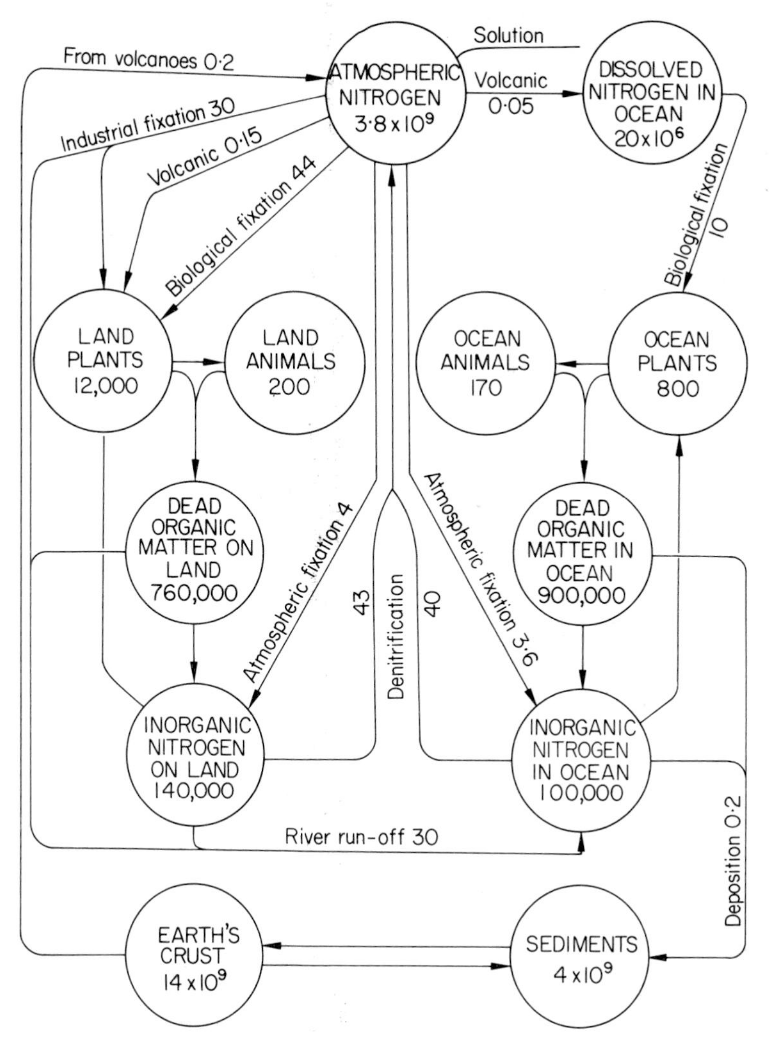

Figure 4.1. The global nitrogen cycle. The units are 10^6 metric tonnes for the pool sizes and 10^6 metric tonnes/year for the flows between pools. The figures are estimates. The amount in the earth's crust and in the sediments is assumed to be constant, though because of the recent industrial fixation there is about 9×10^6 metric tonnes/year more nitrogen fixed than is denitrified: it is possible that most of this is going into extra biomass, but some may be accumulating in sediments. (Data from C.C. Delwiche, 1970. *Scientific American* **223**, 136–146.)

is little ammonium in the surface waters of lakes (about 0.05 to 0.1 mg/l is common), unless there is heavy contamination with organic matter such as sewage, but it accumulates in poorly aerated lower layers (e.g. 0.1 to 1 mg/l) mostly as a result of the decomposition of the organic matter in sediments (Fig. 4.2). The level of nitrate is generally low for it is used by the phytoplankton in the surface layers and is lost by denitrification (see below) in anaerobic conditions. The levels of organic nitrogen also vary with the seasons (Fig. 4.2) and have been mentioned before (Table 3.2, page 22). The organic nitrogen in surface waters accounts for 50 to 75% of the total soluble nitrogen and the majority of this is as amino groups in proteins, peptides and free amino acids. Much of the rest is in the non-amino groups of these same compounds, though others such as urea are also present.

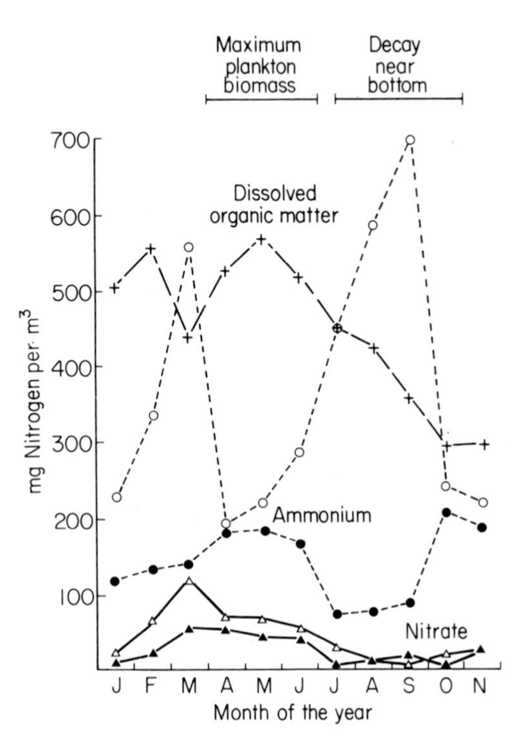

Figure 4.2. The annual variation in the levels of the different forms of nitrogen in Lake Mendota, U.S.A. △——△ nitrate near the bottom; ▲——▲ nitrate near the surface; ○----○ ammonium near the bottom; ●---● ammonium near the surface; +———+ dissolved organic nitrogen (a different year from the other values). Mineral N falls as plankton takes it up and ammonium rises, particularly in deep water as the plankton is decomposed. (Modified from G.E. Hutchinson (1957). *A Treatise on Limnology*. Vol. 1. Wiley, New York.)

The distribution of the forms of nitrogen in the sea (further reading 8) has been studied rather less than in lakes. Most of the open ocean contains about 0.45 mg nitrogen/l; coastal and particularly estuarine waters may contain more. Of this about 95% is as dissolved nitrogen gas and 65% of the rest is as nitrate or nitrite. The nitrate levels increase with depth, and are very low in surface waters in summer when it is taken up by the phytoplankton. At great depths, particularly in the tropics, where anoxic conditions prevail, the nitrate may be reduced to ammonium, as it is in lakes. Ammonium levels in the surface waters vary widely with the season and the levels of plankton.

Soil nitrogen

The amounts and types of nitrogen in the soil vary much more widely than in water depending largely on the organic matter content. The total nitrogen in agricultural soils is generally 0.02 to 0.5% (on a soil dry weight basis) and 40 to 50% of this is in bound amino acids and some amino sugars within the 'humus' material. Soluble forms of nitrogen can be leached out of the soil, for it is not strongly adsorbed or otherwise physically bound to the soil particles; eventually this leached nitrogen enters rivers and lakes.

MICROBIOLOGY OF THE NITROGEN CYCLE

Micro-organisms are of great importance in the nitrogen cycle. Fungi and bacteria are responsible for the decomposition of proteins, chitin, urea etc. and bacteria, particularly, are responsible for the chemical conversions between the various oxidation states of nitrogen including ammonium, nitrogen gas, nitrite and nitrate. The chemical reactions of nitrogen compounds in the biosphere are summarized in Figure 4.3. As far as is known the reactions are similar in both

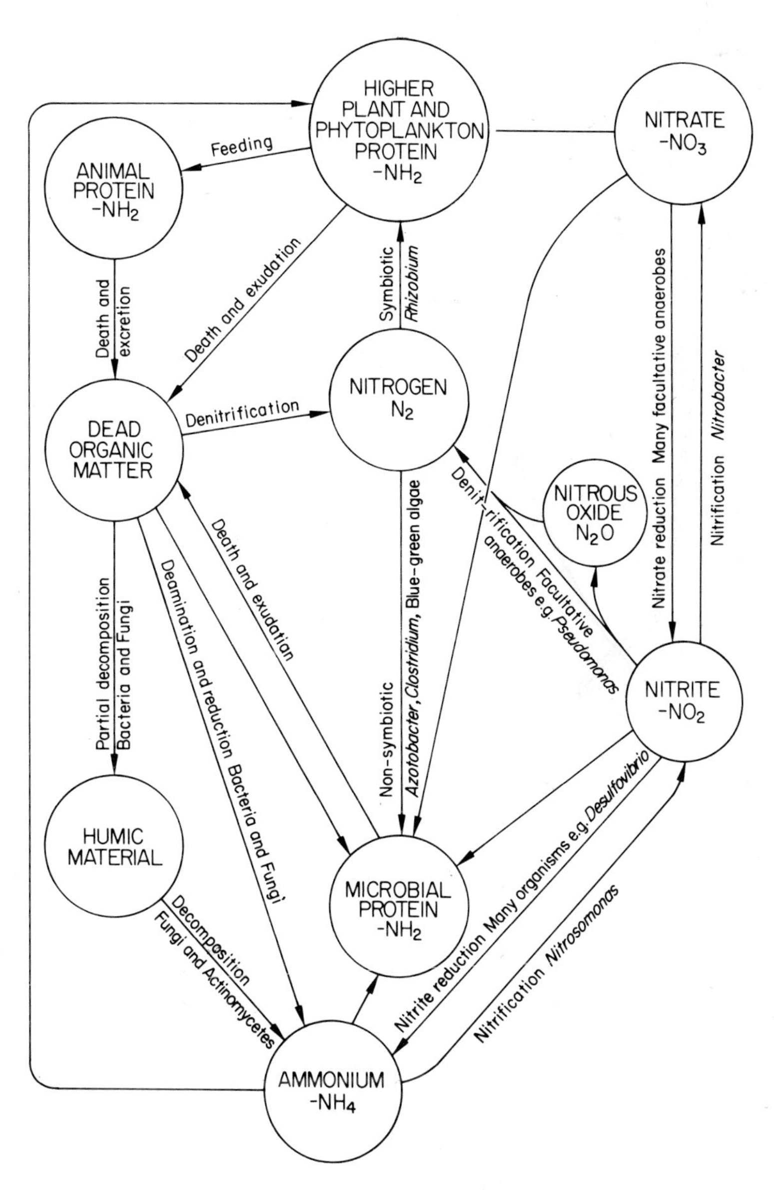

Figure 4.3. The chemical conversions in the nitrogen cycle. The main activities are indicated, together with some of the organisms concerned.

aquatic and terrestrial environments with the exception of the symbiotic association of *Rhizobium*, which is absent from water.

We have considered the decomposition of nitrogen compounds (left-hand side of Fig. 4.3 and Chapter 3) which may be soluble amino acids, small peptides or proteins or insoluble proteins and structural materials such as chitin. Nitrogen is almost always in short supply during decomposition; plant residues have C:N ratios as high as 30:1 and protein is about 5:1. The given amount of nitrogen in the residue is therefore repeatedly recycled through the decomposer microbes; each time some is hydrolysed to amino acids and used by the microbes and some (about 80% in soil) is deaminated to give ammonia and the carbon is respired.

A variable amount of organic nitrogen may be incorporated in peat and other sedimentary deposits or it may be adsorbed onto organic compounds which are not easily decomposed.

Nitrification

The net result of the decay of nitrogenous organic matter is the production of ammonium. This may be taken up by micro-organisms and some higher plants direct, though for the latter nitrate is often the more easily assimilated nitrogen source. Nitrification, which is the conversion of ammonium to nitrate, is carried out entirely by micro-organisms. *Nitrosomonas* converts ammonium to nitrite and *Nitrobacter* converts nitrite to nitrate (further readings 1 and 10). There are reports of various other genera e.g. *Nitrosocystis* and *Nitrosococcus* performing these functions in the sea. The conversion of nitrite to nitrate goes fastest and so nitrite levels are usually low. The organisms are obligate aerobes and are chemoautotrophs (they derive energy from the oxidation and can use this to fix carbon). The conversion of ammonium to nitrite yields 65 to 66 kcals per mole, and nitrite to nitrate about 17 to 18 kcals per mole (Fig. 4.4), which is a low level of energy yield when compared with respiration (686 kcal per mole of glucose) or deamination (176 kcals per mole of ammonium).

The chemistry of nitrification is not clear: some of the enzymes have not been isolated and some of the postulated intermediates cannot be detected (further reading 1). However the first step is apparently oxidation of ammonium to hydroxylamine using molecular oxygen (Fig. 4.4). At the slightly alkaline pH which is optimum for *Nitrosomonas*, ammonia is almost all in the form of the ammonium ion and hydroxylamine is not ionized. Hydroxylamine is toxic but the concentration is usually low, the equilibrium is towards ammonium and the reaction proceeds only as hydroxylamine is oxidized to nitrite. There are intermediates, possibly the nitroxyl radicle and nitrohydroxylamine though whether these exist free is not known. Parts of the flavoprotein and cytochrome system are required for the oxidation of hydroxylamine. *Nitrobacter* then oxidizes the nitrite to nitrate in a single step with molecular oxygen as the terminal electron acceptor mediated by a cytochrome system and generating ATP (Fig. 4.4).

Nitrification by chemoautotrophs occurs in all aerobic environments investigated. It is very slow in the open oceans, being confined to surface water, though in sediments and in estuarine conditions it is easier to demonstrate. In aerobic fresh water nitrification is a well established phenomenon and seems to be the same process, with the same micro-organisms, as has been so extensively studied in soil. Nitrification occurs rapidly near neutral pH, but it is not certain how it proceeds under acid conditions, such as in many moorland and forest soils and waters. *Nitrosomonas* and *Nitrobacter* strains that have so far been isolated do not carry out nitrification much below pH 6 (page 47).

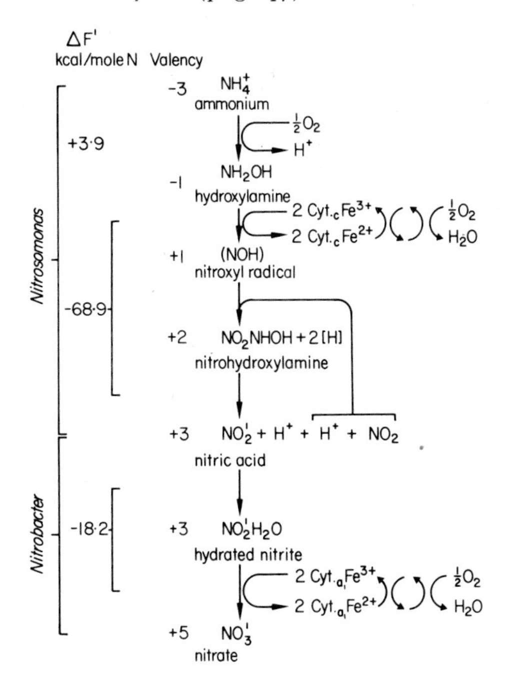

Figure 4.4. Postulated steps in nitrification.

Nitrate reduction and denitrification

The opposite process to nitrification is reduction of nitrate through nitrite to ammonia, and linked with this is denitrification in which nitrate is converted to nitrite then to nitrogen or nitrous oxide, resulting in the loss of gaseous nitrogen to the atmosphere (further reading 6). Assimatory nitrate reduction is carried out intracellularly by those bacteria fungi and blue-green algae that use nitrate and reduce it to the amino group (-NH_2) in proteins. This usually only occurs under aerobic conditions and the enzyme is repressed by free ammonium (cf below, dissimilatory reduction). The chemistry has been little investigated for most micro-organisms but nitrite and hydroxylamine are usually intermediates. In eukaryotes the enzyme contains molybdenum but this has not been detected in the assimilatory nitrate reductase of prokaryotes.

Dissimilatory conversion of nitrate in soil or water is mostly performed by facultative anaerobes which use nitrate as a terminal electron acceptor in the absence of oxygen (e.g. *Thiobacillus denitrificans*, and some species of *Pseudomonas*, *Bacillus*, *Micrococcus* and *Achromobacter*). The process occurs in waterlogged soil, in deep waters or sediments and where the rapid decomposition of organic matter gives rise to anaerobic microhabitats. The reaction is catalysed by a particle-bound enzyme-cofactor complex, known as nitrate reductase, which contains molybdenum (Fig. 4.5) and which is not repressed by ammonium. There is competition between this enzyme and cytochrome oxidase, which is dependent on the relative concentrations of oxygen and nitrate, the electron transport chain is common to the two enzymes as far as cytochrome b but the cytochrome c is different for the two pathways. Oxygen is the preferred electron acceptor in most organisms though some, like *Denitrobacter licheniformis*, have an unusually high affinity for nitrate. From nitrite the process may proceed either to ammonia catalysed by nitrite reductase which seems to require ferredoxin, or to nitrous oxide or nitrogen gas. The mechanisms and the intermediates are

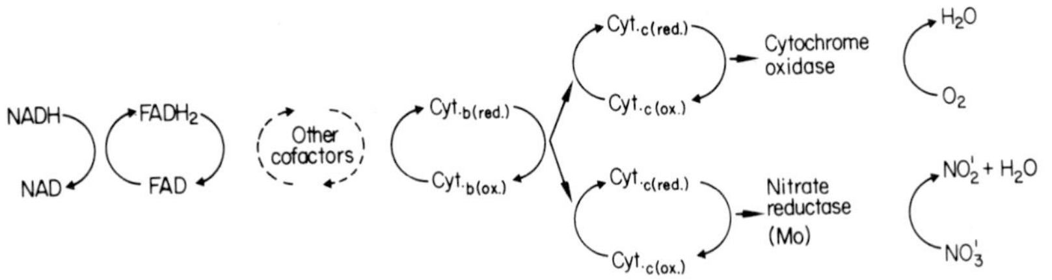

Figure 4.5. The use of nitrate as an electron acceptor in competition with oxygen: this is dissimilatory reduction of nitrate by nitrate reductase which contains molybdenum.

unknown but a possible scheme (Fig. 4.6) has been suggested which does not conflict with the present evidence but for which there is little positive proof.

The loss of nitrogen from soil and water caused by denitrification is important in the world nitrogen balance: it has been estimated that there is an average loss of 22.5 kg/ha/year of nitrogen over the total land surface of the earth. There are no estimates for the sea, but it is probably less per unit area, though the overall total might be greater because of the enormous area of the oceans.

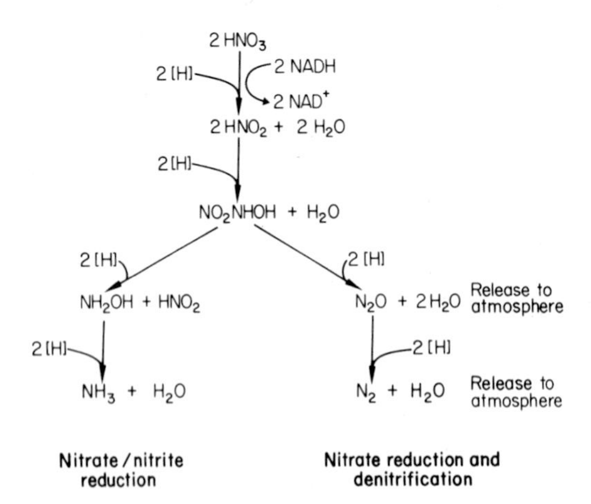

Figure 4.6. A postulated scheme for denitrification. The nitrate reductase contains molybdenum while some, if not all, of the nitrite reductase enzymes contain copper and iron.

Nitrogen fixation

Nitrogen fixation is undoubtedly the most intensively studied aspect of the whole nitrogen cycle (further reading 3, 5, 7 and 9). The ability to perform this is restricted to the prokaryotes, though within this group the capacity is found in diverse organisms. These may be autotrophic or heterotrophic, aerobic or anaerobic, and free living or symbiotic. The process requires anaerobic conditions, or at least reduced oxygen tension, and the aerobes usually have structures to isolate the nitrogen fixation site from the environment. There are no substantiated reports of eukaryotes being able to fix nitrogen.

The photosynthetic bacteria include obligate anaerobes like *Chromatium* and facultative aerobes such as *Rhodospirillum rubrum* and *Rhodopseudomonas palustris*. The amount of nitrogen fixed is thought to be small and economically unimportant, though this group has not been extensively studied (further reading 5 and 9). Blue-green algae (Cyanophyta) are more important, particularly in shallow water and in waterlogged soils such as rice paddies where they are thought to be the main source of nitrogen; repeated crops of rice can be grown without artificial fertilization. It has been claimed that blue-green algae are important in the open ocean, but the amount of nitrogen fixation occurring in the sea has been widely disputed. Nitrogen fixation in blue-green algae is particularly associated with those genera that have heterocysts such as *Anabaena*, *Aphanizomenon* and *Nostoc* but there are also reliable reports of the unicellular *Gloeocapsa* and the filamentous *Plectonema* fixing nitrogen, even though they do not have heterocysts; they do however only operate at low partial pressures of oxygen. Blue-green algae may form lichens (symbiotic associations with fungi), and nitrogen fixation is often more efficient in the association though it does occur in the free-living organism.

The most widely studied free living heterotrophs that fix nitrogen are species of *Azotobacter* (aerobes) and *Clostridium* (anaerobes) particularly *C. pasteurianum*. Nitrogen fixation by some *Klebsiella* spp., *Bacillus* spp., *Mycobacterium* spp., *Desulfovibrio* spp. and an acidophilic tropical genus, *Beijerinckia*, has also been confirmed. There is a very long list of supposedly nitrogen fixing heterotrophs whose credentials are less well authenticated. *Azotobacter* and *Clostridium* are cosmopolitan though their importance in fixing nitrogen in the natural environment remains in some doubt because of the limited energy sources available in most situations to free living heterotrophs. Furthermore, though widely distributed, *Azotobacter* is not normally numerically very important. In most soils nitrogen fixation is also reduced or eliminated by feedback inhibition because of the existing levels of available nitrogen. However there is now evidence that nitrogen fixation by free living organisms may be important in the soil immediately around plant roots (the rhizosphere, Chapter 6). The roots supply carbohydrate as an energy source and maintain low fixed nitrogen levels. Fixation may add up to 39 kg N/ha/year (max. daily rate 9.4 kg/ha) to the soil nitrogen in temperate regions and much higher values are associated with roots of tropical grasses and cereals (page 82).

The symbiotic associations of heterotrophs and higher plants are the major terrestrial source of biologically fixed nitrogen. There are several non-leguminous associations in which fixation has been demonstrated, the best known of which, in temperate regions, is the large nodule formed by alder (*Alnus* spp.) which has been shown to fix nitrogen quite efficiently and to have a significant effect on the nitrogen balance of some forest ecosystems (Fig. 4.7 and Table 4.1) and in some

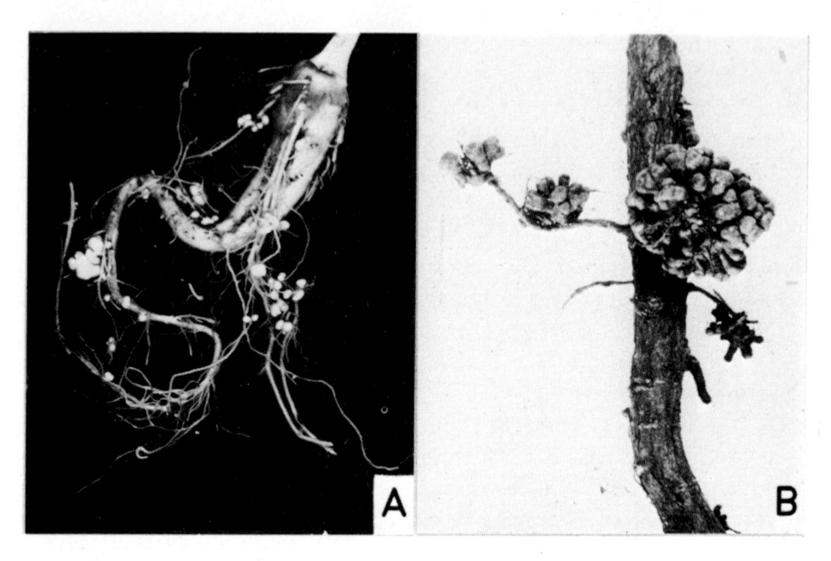

Figure 4.7. Root nodules of A. *Phaseolus coccineus* (½ natural size). B. *Alnus* (natural size).

poor soils such as glacial moraines. The bacterium involved is an actinomycete (*Frankia*), though it has never been isolated in pure culture on artificial medium and successful reinoculation has only been obtained from isolates grown in callus cultures. There are many other non-leguminous plants that have root nodules,

Table 4.1 The rates of nitrogen fixation by different systems. (Selected from R.C. Burns and R.W.F. Hardy (1975) Nitrogen fixation in Bacteria and Higher Plants. *Molecular Biology, Biochemistry and Biophysics* **21**. Springer-Verlag, Berlin. pp. 189).

Nitrogen fixing system	Average Amounts of Nitrogen Fixed kg/ha/year
NON-LEGUMES with nodules	
Alnus spp.	56–156
Acacia sp.	270
Casuarina equisetifolia	58
Myrica gale	9
Hippophae rhamnoides	15–179
LEGUMES with nodules	
Clovers, *Trifolium* spp.	105–220
Lupin, *Lupinus* sp.	150–169
Lucerne, *Medicago sativa*	128–300
Soy bean, *Glycine max*	57– 94
Mixed legumes	125
NON-SYMBIOTIC	
Various agricultural soils	1–90
Forest soils	0–50
Desert crust	2–41
Temperate lakes	0.4– 8
Tropical lake	44

for example *Casuarina* in the tropics, and buckthorn (*Hippophae*) which colonizes sand dunes and gravel soils.

There are leaf nodules in plants of the genera *Ardisia*, *Pavetta* and *Psychotria*, some of which fix nitrogen very slowly and are unlikely to produce significant amounts. The organisms concerned are usually placed in the genera *Chromobacterium, Klebsiella, Phyllobacterium* or *Xanthomonas*. There is a less well defined symbiotic relationship, which seems to be particularly common in the tropics, in which nitrogen fixing organisms e.g. *Beijerinckia* spp. and *Azotobacter* spp., live on the leaf surface (Chapter 8) or in water trapped around the leaf bases of various plants.

The most important and the best studied symbiotic association is undoubtedly that between *Rhizobium* spp. and various Leguminosae (further reading 5). *Rhizobium* occurs free living in the soil but does not fix nitrogen in this situation. There are several species described, usually the main distinction is in the range of host plants which they will infect to produce nodules. There are however many strains within each species which have varying degrees of infectiveness for different host strains and also varying degrees of efficiency of nitrogen fixation in the symbiotic association. These two variables are not linked and it is possible to get both highly infective strains that are not efficient fixers and also nitrogen fixing strains that are not infective when in competition with other rhizobia. It is therefore important to ensure that compatible, effective strains exist in the soil, and if they do not then it can be inoculated with the correct *Rhizobium* strain (Table 4.2). The inoculum needs to be massive and of an aggressive strain if it is to be successful, for the resident microflora resists changes (Chapters 1 and 6). The scale of nitrogen fixation by leguminous plants is difficult to estimate and

Table 4.2 The effect on the crop yield of inoculation of seed with the appropriate *Rhizobium* strain. The increases were statistically different from the uninoculated controls. (Modified from E.N. Mishustin and V.K. Shil'nikova (1971) *Biological Fixation of Atmospheric Nitrogen*. Macmillan, London, pp. 420.)

Plant	Crop Parameter	Percent increase in weight at harvest
Pea, *Pisum* sp.	Foliage	23.5
	Seed	14.9
Broad bean, *Vicia faba*	Foliage	16.6
	Seed	18.0
Lupin, *Lupinus* sp.	Foliage	66.1
	Seed	20.9
Soy bean, *Glycine max*	Foliage	19.1
	Seed	12.3*
Lucerne, *Medicago sativa*	Hay	26.9
	Seed	57.9
Sainfoin, *Onobrychis viciifolia*	Hay	31.1
	Seed	46.6

*The seed may also have an increased protein content, about 16% more than the uninoculated plants on a dry weight basis.

varies with the particular crop and the soil conditions, but values between 50 and 300 kg/ha/year are common (Table 4.1).

The process of root nodule formation is as follows. The bacteria in the soil produce the plant growth hormone indole acetic acid (probably by metabolizing tryphtophan which is exuded by the plant root) and also a root hair curling factor, which causes the root hairs to become twisted and deformed. The wall of the root hair invaginates and grows backwards down within the root hair forming an infection thread containing the bacteria. This infection thread grows across the root, passing through the cortical cells until it reaches polyploid cells which are normally present in small numbers at the points of potential lateral root initiation. The infection thread bursts and releases the bacteria into the cytoplasm of the polyploid cells. The bacteria start to divide, they lose their cell walls, become swollen and often branch in an 'X' or 'Y' shape; in this form they are called bacteroids and they are enclosed individually or in small numbers within the folds of the host plasma membrane. It is this combination of bacteroids and polyploid cells that actively fixes nitrogen. The cells divide rapidly, forming the core of the nodule, and the surrounding diploid cells also divide and differentiate to cover the nodule in cortical tissue and to form vascular connections with the root.

The biochemistry of nitrogen fixation is remarkably similar in all the organisms that have been studied (further reading 3 and 7). The enzyme system involved, called nitrogenase, requires molybdenum and magnesium, and cobalt in root nodules. The first detectable product is ammonia. The enzyme is extremely sensitive to oxygen, which is hardly surprising considering the strong reduction involved in the conversion of nitrogen to ammonium. Nitrogen fixation in aerobes such as *Azobacter* was at first thought not to be oxygen sensitive, but the enzyme is carried on subcellular particles which protect the system from oxygen: the purified enzymes are just as sensitive as those derived from anaerobes. This sensitivity greatly complicates the studies of the system for the extraction and purification has to be performed under completely anaerobic conditions. The energy requirement for the reaction is high, 10 to 20 molecules of ATP for every molecule of nitrogen fixed (this is equivalent to about 1 g of glucose per 20 mg of nitrogen). The reason for the high energy demand is not clear, for the overall reaction of nitrogen to ammonium is just exothermic and should therefore need no energy, given a perfect catalyst.

Nitrogenase has two main components: a protein (protein 1) containing molybdenum and iron with a molecular weight of about 200,000 and a smaller protein (protein 2) containing only iron. These two fractions must act together for there to be any nitrogen fixation. Ferredoxin, a non-haem iron protein with the very low redox potential of -0.417, is the electron donor to nitrogenase in *Clostridium* and some other organisms, but different sources of reducing power (e.g. flavodoxin) has been described from other systems. ATP is produced by oxidative phosphorylation (Fig. 4.8) or from pyruvate. Photosynthetic organisms may supply some of the ATP from cyclic phosphorylation (further reading 5 and 9). Various coenzymes, the cytochrome system and other metabolic pathways can therefore have a role in providing ATP and reducing power but they are not themselves central to the action of nitrogenase. In experimental systems

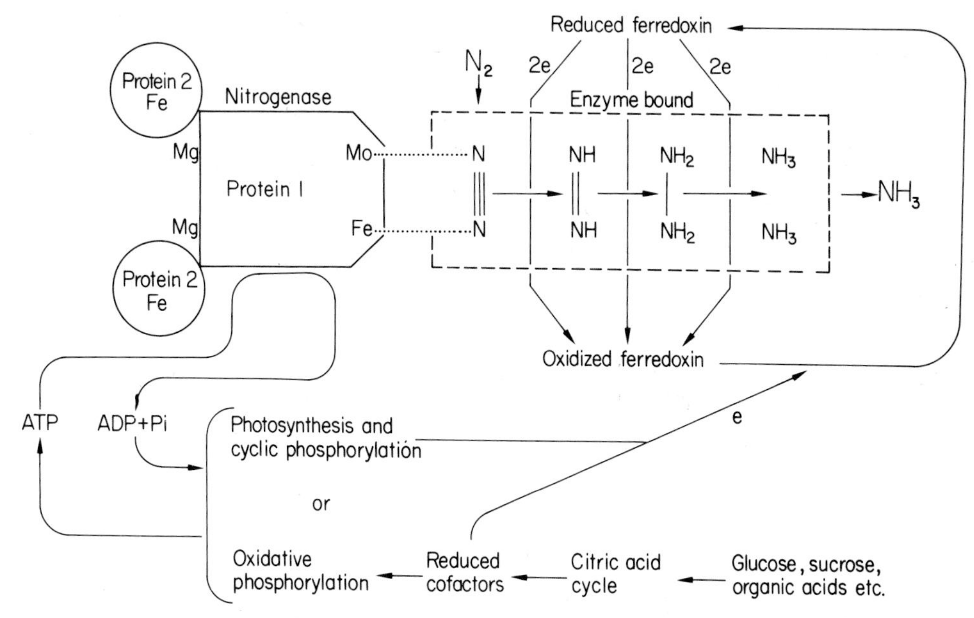

Figure 4.8. A generalized scheme for the action of nitrogenase, incorporating results from studies on various organisms.

pyruvate may fulfil both functions or creatine phosphate can be used as part of an ATP generating system with reducing power from such inorganic compounds as sodium dithionite.

Nitrogenase is not very specific and will catalyse a number of reductions, particularly of triple bonded compounds such as acetylene, cyanide and azide. This has proved most useful and acetylene reduction has become the standard test for nitrogenase: ^{14}C acetylene can thus be used instead of ^{15}N or the ethylene produced from the acetylene can be detected by gas liquid chromatography. The active site on the enzyme was long disputed since nitrogen was so chemically inert. It is now agreed that the nitrogen binds to the molybdenum-iron protein. Such a complex now seems plausible in natural systems, because compounds of nitrogen and metals, particularly cobalt, have recently been synthesized chemically. We now come to the usual problem with inorganic nitrogen metabolism of just what intermediates are involved: di-imide (HN–NH), hydrazine $H_2N–NH_2$), and hydroxylamine (NH_2OH) have been suggested but they are toxic, unstable at physiological pH etc. and they have never been detected. Nitrogen remains enzyme-bound as it is reduced in two-electron steps to ammonium, the latter being released as the last N–N bond is broken (Fig. 4.8).

Of the symbiotic organisms only *Rhizobium* has been studied in any detail. The effective nodules are a pink colour and this is caused by the presence of a pigment, leghaemoglobin, which occurs only in active nodules and not in either symbiont or host growing alone, or in ineffective nodules. The purpose of the leghaemoglobin has been much disputed but the present theory is that it acts as an oxygen carrier and reduces the oxygen tension. The nitrogenase system is in the bacteroids. The ATP generating system and/or the reducing power generating system have from time to time both been allocated to the higher plant. More recently it has been shown that both are in the bacteroids, which use substrates supplied by the host. If all the necessary systems are in the bacteroid the question arises as to why the free living *Rhizobium* (the bacterium not bacteroid) in soil does not fix nitrogen. It may be simply the supply of substrates or it may be connected with the leghaemoglobin. There are differences in the cytochrome systems between cultured cells and the bacteroids and also in the ability to oxidize some substrates. It has recently been demonstrated that some cultured *Rhizobium* strains (the bacterium) will fix nitrogen if supplied with carboxylic acids, pentose sugars (arabinose or xylose) and also small amounts of fixed nitrogen. This latter point is a puzzle, for available fixed nitrogen tends to depress fixation by the nodules. There is evidently much more to be learnt about the interaction between *Rhizobium* and its host.

THE EFFECT ON THE NITROGEN CYCLE OF THE ENVIRONMENT AND AGRICULTURAL PRACTICE

The direction and rates of flow in the nitrogen cycle are of course affected by environmental factors. If anything the effects are more noticeable than in the carbon cycle because the latter has a multitude of different micro-organisms which between them can cope with almost any conditions. The nitrogen cycle on the other hand is dependent on a few different genera of bacteria for fixation and even a single genus for particular conversions e.g. *Nitrosomonas* and *Nitrobacter*. The main factors controlling the nitrogen cycle are the general ones of temperature and water availability. Under aerobic conditions nitrification and symbiotic nitrogen fixation can occur; in anaerobic conditions denitrification occurs and nitrogen fixation only by *Clostridium* and some photosynthetic and sulphate reducing bacteria. In normal agricultural soils the balance between nitrification, nitrogen fixation and denitrification is determined by aeration. The processes can take place in the same soil because of the presence of relatively aerobic and anaerobic microhabitats.

In some situations it might be desirable to encourage denitrification. After normal, or even the best, treatment of sewage the effluent can contain high levels of inorganic nitrogen, even though the organic matter content has been reduced to a satisfactory level. Nitrogen can give trouble because it encourages algal growth downstream of the outflow (Chapter 7). Treatment plants to encourage nitrification to nitrate followed by denitrification to gaseous nitrogen are theoretically possible and they have been used on a pilot scale by manipulating the trickle filters or activated sludge tanks and the anaerobic sludge digesters to give an aerobic then a relatively anaerobic phase in the treatment. The commercial use of such systems is limited, at the moment, by the cost of the extra facilities, but the biological removal of nitrogen from effluent is possible if or when more stringent standards are imposed on effluents.

Recently there has been a great increase in the amounts of ammonium or nitrate fertilizers added to soil to increase the yield of many important crops. Ammonium tends to make the soil acid, particularly when applied as ammonium sulphate, and nitrate tends to give alkaline conditions. The ammonium is adsorbed on soil colloids and within clay particles because of its positive charge, but the nitrate is very mobile and tends to be more easily leached out. Heavy modern machinery may compact wet ground and give anaerobic conditions, which encourages denitrification.

The fate of nitrogen in fertilizer is dependent not only on the soil conditions but also on the chemical nature of the fertilizer, its rate of application and the granule size. Thus a substance like urea, which hydrolyses to give an alkaline environment, can be applied at high enough rates or in such large granules as to raise the soil pH appreciably. Nitrification is favoured at neutral or slightly alkaline pH but above pH 8 *Nitrobacter* does not convert much nitrite to nitrate. *Nitrobacter* is also more sensitive to ammonium toxicity than *Nitrosomonas*. The raising of the soil pH in the vicinity of fertilizer granules and the excess of ammonium can lead to a lack of nitrification, or to an accumulation of nitrite which is phytotoxic. On the other hand nitrification will not occur if the soil is too acid and liming soils with a pH between 4 and 6 increases nitrification.

The level of ammonium in the soil also influences nitrification since it is the initial substrate: it may be used very rapidly in some soils e.g. grasslands, and so not be in sufficient quantities for nitrification to proceed. There are also reports that grasses can inhibit *Nitrosomonas* and *Nitrobacter* by their root exudates. Grasslands therefore tend to have very low levels of nitrification.

It is important for fertilizers to be formulated and applied so that microbial nitrification is controlled to get a slow, steady release of nitrate which the plants can take up before it is leached or denitrified. Care must also be taken not to inhibit beneficial nitrogen fixation. The application of fertilizer is not a simple task.

Various other agricultural practices can also affect the conversion of nitrogen in the soil. For example, many herbicides inhibit nitrite reduction in green plants, and herbicidal properties have even been attributed in part to the toxic effect of nitrite accumulation when nitrate is the plant's nitrogen source. Some of the chlorinated hydrocarbon insecticides, and herbicides like 2.4-D inhibit microbial nitrification in laboratory studies at levels that are within the possible concentrations in the soil. Significant effects in the soil are not however very common, possibly because some of the pesticides are adsorbed onto soil colloids.

Because of the high cost of fertilizers, and the high level of industrialization needed to produce them, there has been a lot of work to try to manipulate the soil microflora to produce more fixed nitrogen, and to conserve that already present by discouraging denitrification. It is thus hoped that yield increases can be obtained without the reliance on artificial fertilizers. The most obvious possibility is to add nitrogen fixing organisms to the soil. There is the case of *Rhizobium* strains, discussed above, which is well substantiated and of proved usefulness. Rather more controversial has been the attempt to inoculate soil or seeds with *Azotobacter*. This work was mainly carried out in the USSR and claims of significant increases in yield were made for the large scale application on Russian farms. Up to 20% increases were claimed for a wide variety of important crops and this was assumed to be due to nitrogen fixation. It has been pointed out already that the energy requirements of *Azotobacter* fixing nitrogen are high and calculations suggested that there was insufficient organic matter in the soil to provide any reasonable quantity of utilizable nitrogen. There was controversy for many years but eventually many of the yield increases have been substantiated in all parts of the world, particularly when something less than the best seed was used. It turns out that many micro-organisms can give yield increases while being unable to fix nitrogen. The most likely explanation is that the added micro-organisms, including *Azotobacter*, are giving protection to the plant from various pathogenic organisms or are producing seed germination stimulants and growth promoting substances such as gibberellins (Chapter 6).

The form in which nitrogen is applied to the soil can affect plant pathogens, particularly soil-borne root diseases. There is no general rule: alkaline soils and nitrate fertilizers favour some pathogens while acid soils and ammonium fertilizers favour others. The effect is probably not directly on the pathogenic micro-organisms, for the pH changes are small and most probably the change in disease severity results from many small changes in the competitive balance of the soil (Chapter 6).

In the industrial countries with intensive agriculture we think, rightly, that fertilizers and leguminous root nodules are very important as suppliers of nitrogen. In most of the world's agriculture, and in all natural habitats, the great majority of the nitrogen is supplied by the mineralization of organic matter to ammonium and ultimately to nitrate. The importance of this part of the nitrogen cycle, where the decay is brought about by many different micro-organisms, is often overlooked amongst the impressive discussions of nitrogen fixation which in these situations merely serves to top-up the nitrogen supply, replacing losses by leaching and denitrification. As mentioned above, during mineralization some of the nitrogen is immobilized as microbial proteins and structural polymers. The nitrogen content of ploughed-in crop residues must be greater than about 2% for there to be an immediate release of nitrogen, below this no nitrogen will be released, indeed available nitrogen in the soil might also be immobilized by the micro-organisms. Over a period of time, as the nitrogen is re-used by the respiring organisms the carbon content of the decomposing material will drop and the nitrogen content therefore rises until it reaches the level where there is net release of inorganic nitrogen. Regardless of the amount of nitrogen in the substrate the processes of mineralization and immobilization both occur, but at low nitrogen levels the immobilization is greater. Immobilization is not necessarily detrimental for it reduces the amount lost by leaching and retains nitrogen in the soil for later release and use by micro-organisms and higher plants.

The discussion up to now has been mainly concerned with soils and with agriculture. Compared with the land there is very little known about the nitrogen cycle in aquatic environments. It is assumed that the same environmental factors control the rates of the various processes and the direction of flow of such opposed reactions as nitrification and denitrification. Individual lakes have been studied in some detail and they show seasonal variations as might be expected (Fig. 4.2) due to fluctuations in plankton population. Most water temperatures, particularly in the deep oceans, are low and the rates of nitrogen conversions are therefore slow. One of the main controlling factors is again the degree of aeration, and oxygen tensions can be very low in deep water or in sediments rich in organic matter. In most productive lakes it is probable that denitrification is important;

nitrogen balances show a net loss of nitrogen in the passage of water through the lake (Fig. 4.9): some of this must end up in the sediments but most is probably lost as gaseous nitrogen. In less productive lakes nitrogen fixation about balances or slightly exceeds denitrification (further reading 4). The nitrogen imbalance is also a feature of the marine systems that have been studied. Again the excess nitrogen which enters the system, but does not appear to accumulate within it or to leave it, is probably lost by denitrification. There are very few measurements of the rates of *in situ* denitrification in the sea, but denitrifying organisms occur and conditions are suitable, particularly at moderate and great depths.

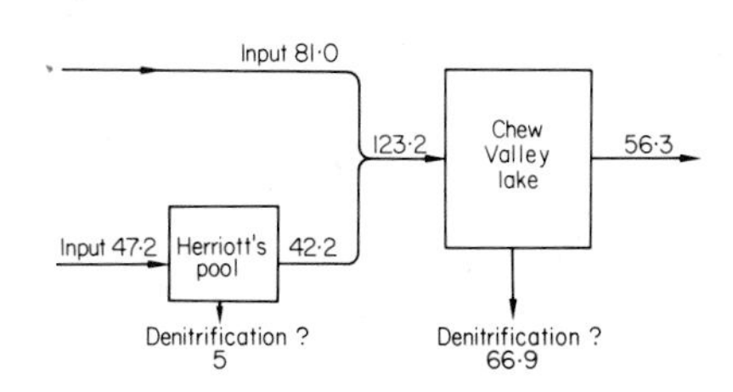

Figure 4.9. The estimated nitrogen budget in 1969 for a river system, including a small pool and a large artificial lake, near Bristol, England. Figures are in metric tonnes of nitrogen (total nitrogen). (Data from R.S. Wilson *et al.* (1971). *The Limnology of Chew Valley and Blagdon Lakes*. Department of Zoology, University of Bristol: calculated from data supplied by Bristol Waterworks Company.)

Most of the nitrogen fixation in lakes, and probably also in the ocean, is by blue-green algae and it is usually light dependent and inhibited by nitrate. In summer and autumn the available nitrogen levels are low and organic matter (for an energy source) from previous growth of algae could result in favourable conditions. Though the amount fixed is not great (e.g. less than 1% of the total nitrogen input for Lake Windermere in the English Lake District), it may be important if it comes at a time when the other sources of nitrogen are at a minimum. This is a good example of negative feedback control of the nitrogen cycle: as the level of available nitrate nitrogen falls the nitrogen fixation by blue-greens 'switches on' and much of the fixed nitrogen is released as extra-cellular products for use by the other algae.

FURTHER READING

1 Aleem M.I.H. (1970) Oxidation of inorganic nitrogen compounds. *Ann. Review of Plant Physiology*. **21**, 67–90.
Review of nitrification.

2 Brown M.E. (1974) Seed and root bacterization. *Annual Review Phytopathology*, **12**, 181–197.
Includes inoculation studies with Rhizobium.

3 Burns R.C. & Hardy R.W.F. (1975) Nitrogen Fixation in Bacteria and Higher Plants. *Molecular Biology, Biochemistry and Biophysics*, **21**. Springer-Verlag, Berlin. pp. 189.
A comprehensive and recent review of the chemistry and ecological aspects.

4 Hutchinson G.E. (1957) *A Treatise on Limnology*. Vol. 1. Wiley, New York. pp. 1015.
Includes nitrogen distribution in fresh waters.

5 Lie T.A. & Mulder E.G. Eds. (1971) Biological nitrogen fixation in natural and agricultural soils. Special volume; *Plant and Soil*. pp. 250.
A comprehensive collection of reviews, particularly on *Rhizobium*.

6 Payne W.J. (1973) Reduction of nitrogenous oxides by micro-organisms. *Bacteriological Reviews* **37**, 409–52.
Review of denitrification.

7 Postgate J.R. (1971) *The Chemistry and Biochemistry of Nitrogen Fixation.* Plenum Publishing, London. pp. 326.
A comprehensive text.
8 Riley J.P. & Skirrow F. Eds. (1965) *Chemical Oceanography.* Vol. 2. Academic Press, London and New York. pp. 712.
Includes nitrogen distribution in the sea.
9 Stewart W.D.P. (1973) Nitrogen fixation by photosynthetic micro-organisms. *Annual Review of Microbiology*, **27**, 283–316.
Reviews fixation by Cyanophyta.
10 Wallace W. & Nicholas D.J.D. (1969) The biochemistry of nitrifying micro-organisms. *Biological Reviews*, **44**, 359–391.
A review of nitrification.

5 Microbial conversions of other elements in the environment

There are mineral elements, such as phosphorus, calcium and potassium, which play important roles in the growth and metabolism of all organisms. Micro-organisms contain them, and therefore immobilize them, and they also release these minerals as a consequence of the decay of organic matter. The products of microbial metabolism, by altering pH for example, may affect the availability of an ion by altering its solubility or its oxidation state, though that ionic species is not itself metabolized. These are indirect effects on the availability of many plant nutrients in which micro-organisms are probably important, though there is little detailed information. In contrast sulphur and iron have groups of bacteria which specifically use them as substrates. In this chapter we will discuss the microbial transformations of sulphur and iron and also the part played by microbes in the cycling of phosphorus. Silicon is also considered briefly because it is a major constituent of the walls of diatoms, a large and important group of algae.

SULPHUR TRANSFORMATIONS

Sulphur is one of the elements whose distribution has been greatly affected by man in the last 50 or 100 years, mostly by the burning of fossil fuels (Chapter 3) which releases sulphur dioxide into the air. This is washed out by rain and so reaches the soil and water (Fig. 5.1). The effects may be serious on a local scale in, or downwind of, industrial areas where the amount of sulphur involved is considerable (Fig. 5.1).

In soils 10 to 25% of the sulphur is as sulphate which may be readily leached unless adsorbed onto soil colloids. A large part of the remaining 75% is as organic matter from which sulphur is slowly released during decomposition. In anaerobic soils or microhabitats sulphides and polysulphides are frequently present. The total amount of sulphur in soils is usually 100 to 500 mg/kg of which half may be soluble. Sulphur may be a limiting nutrient in intensive agriculture, though rarely in natural environments.

The sulphur in the air is mostly sulphate, sulphur dioxide and hydrogen sulphide: the concentrations present are variable but unpolluted air contains less than 1 to as much as 5 ng/l of each of the three main forms, though this may rise to 20 ng/l in even mildly polluted air. Values up to 200 ng/l of sulphur dioxide have been recorded in city centres with even greater concentrations near particular industrial sources. Rainwater, passing through relatively clean air may collect 2 mg of sulphur/l, but again the total amounts vary: in an unpolluted region with low rainfall (Africa) it may amount to only 1 kg/ha/annum but in industrial Europe and North America it may be as high as 100 kg/ha/annum.

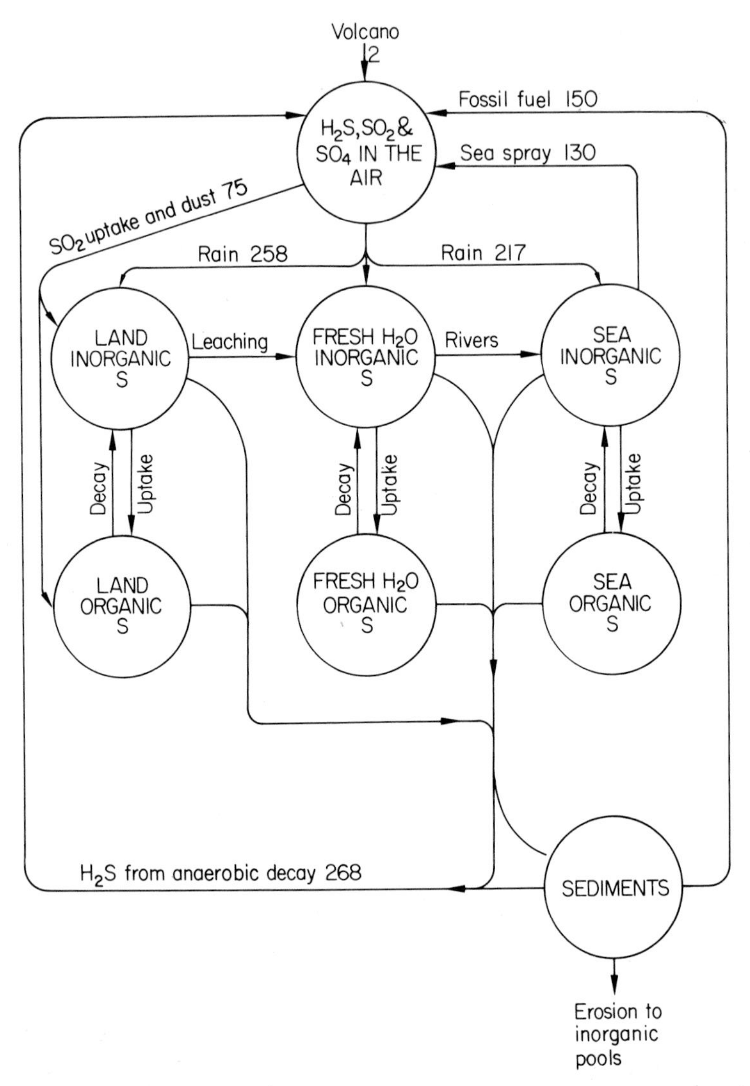

Figure 5.1. The cycle of sulphur in the biosphere. The flow rates between pools are as 10^6 tonnes (as sulphate)/year. (Data from W.W. Kellogg, R.D. Cadle, E.R. Allen, A.L. Lazarus and E.A. Martell (1972). The sulphur cycle. *Science* **175**, 587–596.)

In addition sulphur dioxide may make the rainwater acid, and eventually cause pH changes in soil and water.

In fresh water (further reading 6) the main sources of sulphur are rainwater and the erosion or leaching of the land; usually the concentration is 3 to 30 mg sulphur/l. The sea is the richest source of sulphur, containing up to 2700 mg/l.

Sulphur may exist in many oxidation states, and micro-organisms can carry out the conversions between them (Fig. 5.2). Elemental sulphur, sulphide, sulphur in amino acids and sulphate are the commonest forms. Little is known of the biochemistry and enzymology of the various reactions, but most sulphur

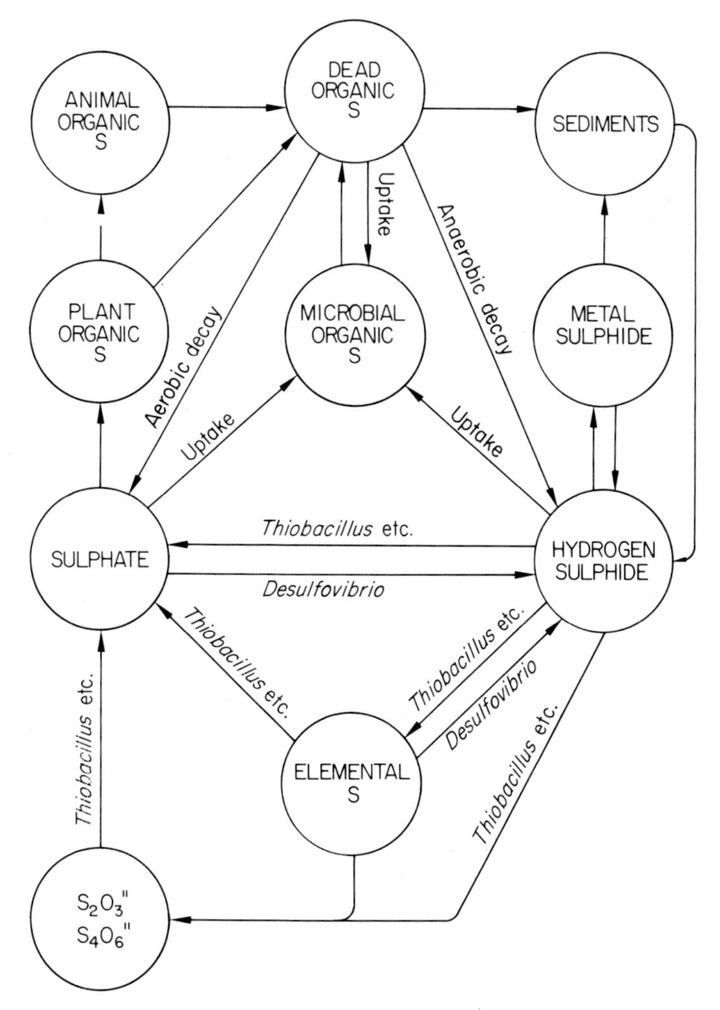

Figure 5.2. The chemical transformations os sulphur compounds carried out by micro-organisms: see text for discussion of the organisms concerned in each reaction. '*Thiobacillus* etc.' refers to *Thiobacillus, Beggiatoa, Thiothrix, Chlorobium, Chromatium,* depending on the light and aeration conditions.

conversions seem to involve adenosine-5′-phosphosulphate (APS) and parts of the cytochrome system (cytochrome c and c_3) with NAD, NADP or ferredoxin. Reduction also involves 3′-phosphoadenosine-5-phosphosulphate (PAPS) (further reading 4). Possible reactions may be summarized as follows:

Assimilatory reduction (micro-organisms, higher plants):

$$\begin{aligned} ATP + SO_4'' &\rightleftharpoons APS + PO_4''' \\ ATP + APS &\rightleftharpoons PAPS + ADP \\ PAPS + NAD(P)H &\rightleftharpoons PAP + SO_3'' + NAD(P) \\ NADH^+ + SO_3'' &\rightleftharpoons SH' + NAD \end{aligned}$$

Dissimilatory reduction (bacteria):

$$ATP + SO_4 \rightleftharpoons APS + PO_4'''$$
$$APS + 2e' \rightleftharpoons SO_3'' + AMP$$

Oxidation (bacteria):

$$2AMP + 2SO_3'' \rightleftharpoons 2APS + 4e'$$
$$2APS + PO_4''' \rightleftharpoons 2ADP + 2SO_4''$$

Considering first the organic sulphur (Fig. 5.2); this is metabolized during decomposition of protein by many sorts of microbes (Chapter 3) and may be released as either sulphate or hydrogen sulphide depending on the aeration conditions. Some of the sulphur is used by the micro-organisms and this immobilized, though sulphur is very rarely limiting during decomposition (cf. nitrogen, Chapter 4) and C:S ratios of up to 200:1 may occur with no limitation of the respiration rate of the micro-organisms causing the decay.

The hydrogen sulphide produced by anaerobic decay (and from sulphate reduction, see below) may precipitate metal sulphides, which are the reason for the black colour of many anaerobic muds. Sulphides may be oxidized by chemosynthetic bacteria if aerobic conditions (positive redox potential, Eh) occur, or by photosynthetic bacteria under continued anaerobiosis. Many heterotrophic bacteria (including actinomycetes) and fungi can oxidize hydrogen sulphide, but the most important organisms in soil and water are the chemosynthetic *Thiobacillus* spp. which can oxidize elemental sulphur, sulphides, thiosulphate or tetrathionate to sulphate. Some can tolerate the very acid conditions which are produced, for example $pH < 1$. Most thiobacilli therefore occupy a particular microhabitat in a region of overlapping concentration gradients where there is oxygen but close enough to anaerobic regions for some reduced form of sulphur, usually hydrogen sulphide, to be present. Such conditions are found in some soils, or in water lying above muds rich in organic matter. One species, *Thiobacillus denitrificans*, is a facultative anaerobe (can live in environments with low or even negative Eh), which can use nitrate instead of oxygen as the terminal electron acceptor (Chapter 4; this is a good example of the linking of the different nutrient cycles). Some of the possible oxidation reactions are given in Table 5.1, but notice that *Thiobacillus thioparus* can oxidize substrates such as thiosulphate or tetrathionate to sulphate while at the same time depositing the reduced form, elemental sulphur. *Beggiatoa* and *Thiotrix* spp., filamentous bacteria that mostly live on the surface of sediments containing hydrogen sulphide, also deposit elemental sulphur, though they are probably mostly heterotrophic rather than chemosynthetic. The free energy changes in sulphur oxidations are not large, being about 40 kcals for sulphide to sulphur and a further 120 kcals to sulphate.

The activities of *Thiobacillus* can have important effects on the pH of the environment, and this has been particularly studied in soils. The acid produced can increase the solubility, and therefore the availability of other elements. Increases in soil acidity have been noted on reclaimed land such as the Dutch polders, where the anaerobic estuarine muds contained sulphides or polysulphides which were oxidized by *Thiobacillus* as the muds dried out and became aerobic. Such pH changes can decrease the flocculation of clay particles and hence worsen the soil structure: lime may be added to correct the pH drift and to

Table 5.1 Examples of the possible sulphur conversions carried out by micro-organisms.

Initial substrate	Organisms	Required Eh, aerobic +, anaerobic −, with the optimum value where known	Possible reactions
Oxidations			
S°	*Thiobacillus thiooxidans* and *T. thioparus*	+580	$2S+3O_2+H_2O \rightarrow 2H_2SO_4$
S_2O_3''	*T. thiooxidans*	+	$S_2O_3''+2O_2+H_2O \rightarrow 2HSO_4'$
S_2O_3''	*T. thioparus*	+	$5S_2O_3''+H_2O+4O_2 \rightarrow 5SO_4''+H_2SO_4 +4S$
S_4O_6''	*T. thioparus*	+	$S_4O_6''+2CO_3'+2O_2 \rightarrow 3SO_4''+S+2CO_2$
S°	*T. denitrificans*	+600 to 0	$6NO_3'+5S+2CaCO_3 \rightarrow 3SO_4''+2CaSO_4 +2CO_2+N_2$
S″	*Thiobacillus spp.* *Beggiatoa spp.* *Thiothrix spp.*	+	$4H_2S+O_2 \quad 4S+2H_2O$
S″	Photosynthetic S bacteria	−	$CO_2+2H_2S \xrightarrow{\text{light}} (CH_2O)_n+2S+H_2O$
Metal S″	*T. ferrooxidans*	+	$2FeS_2+7O_2+2H_2O \rightarrow 2FeSO_4+2H_2SO_4$
Reduction			
SO_4''	*Desulfovibrio sp.*	−180	$SO_4''+4H_2 \rightarrow S''+4H_2O$

reflocculate the clay. On the other hand the application of elemental sulphur, to be oxidized in the soil, has been used agriculturally to lower pH.

The photosynthetic sulphur bacteria, either green such as *Chlorobium* or purple coloured e.g. *Chromatium*, are of importance only in shallow waters with a high organic matter content. They are one of the few means by which reduced sulphur may be oxidized in anaerobic conditions. Most photosynthetic sulphur bacteria therefore occupy a different microhabitat from *Thiobacillus*, one where the water is shallow enough for there to be sufficient light and yet deep enough for anaerobic conditions (Fig. 5.3). However, several species of purple sulphur

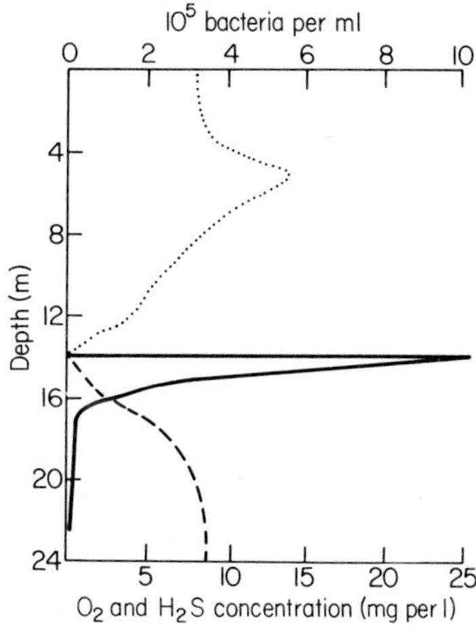

Figure 5.3. The distribution of *Chromatium* sp. (———) in relation to depth, oxygen (.....) and hydrogen sulphide (– – – –) concentration in a lake rich in organic matter. (Modified from Table 93, S.I. Kusnetzov (1959). *Die Rolle der Mikroorganismen im Stuffkreislauf der Seen*. VEB Deutscher Verlag der Wissenschaften, Berlin, pp. 301.)

bacteria are facultative aerobic organotrophs and some can be aerobic chemosynthetic organisms using the oxidation of hydrogen sulphide to sulphate to fix carbon dioxide.

The assimilatory reduction (page 53) of sulphate in the methionine cycle is carried out by all plants and micro-organisms that take up sulphate and use it as $-SH_2$ in amino acids. The dissimilatory reduction of sulphate (and sulphite, thiosulphate and tetrathionate) is a much less common ability: almost all the work has been done on one species of bacterium, *Desulfovibrio desulfuricans* which is an obligate anaerobe with an optimum pH near neutral. The ability of *Clostridium nitrificans* to reduce sulphate has also been investigated, though in much less detail. Both species produce hydrogen sulphide in the presence of sulphate and organic matter, that is they can use sulphate as a terminal electron acceptor. This is similar to the reduction of nitrate (page 40) but the organisms that use nitrate are facultative anaerobes whereas *Desulfovibrio* uses no other terminal acceptors.

The hydrogen sulphide produced by sulphate reduction and by anaerobic protein decay can have serious effects. Firstly it is toxic to most aerobic organisms, including crop plants. Secondly *Desulfovibrio* is very important in metal corrosion; in the sea the sulphate concentrations are high and polluted dock and coastal waters have a high organic matter content and are often anoxic. Metal pipes buried in soil rich in organic matter are also corroded. The sulphide produced by *Desulfovibrio* removes the Fe^{3+} (or other metal) from the metal surface and precipitates it as FeS or Fe_2S_3 and thus causes continued ionization (anodic depolarization). At the same time the hydrogen is removed from the cathodic regions (Table 5.1) and this causes cathodic depolarization. There are many methods of reducing corrosion including the use of biocides, wasting anodes and imposed electrical currents to give cathodic protection (further reading 9).

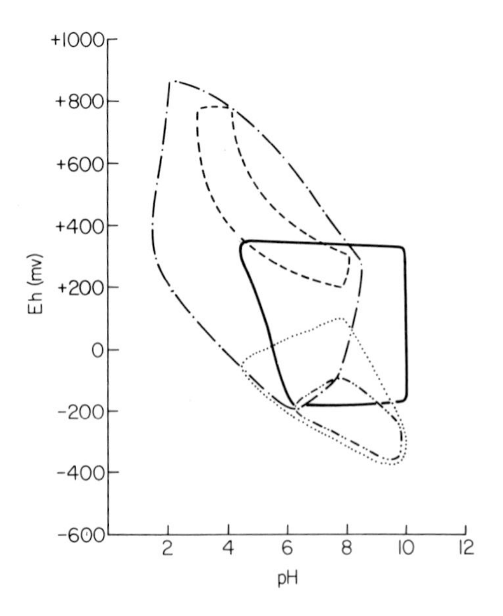

Figure 5.4. The range of bacterial groups concerned with sulphur conversions in relation to pH and Eh. *Thiobacillus* —·—·; Iron bacteria – – – – –; Purple sulphur bacteria ———; *Desulfovibrio* . . . . .; Green sulphur bacteria (*Chlorobium*) —··—··— (Modified from E.J.F. Wood (1965). *Marine Microbial Ecology*. Chapman and Hall, London. pp. 243.)

The organisms concerned with sulphur oxidations and reductions are controlled mainly by aeration (i.e. the redox potential) and by pH. Thus thiobacilli require high redox potentials but withstand low pH, though species vary in their range. On the other hand the purple sulphur bacteria have a very wide pH range, 4.8 to 10.5. We are not dealing with different groups of micro-organisms occupying discrete microhabitats but with groups overlapping each other on gradients of aeration and acidity (Fig. 5.4). A slight change in pH or Eh will change the relative proportions of the species in a mixed population.

The importance of the sulphur bacteria lies not only in the reactions which they carry out in the sulphur cycle but also in the effects of their toxic and acidic end products and in the precipitation of metal sulphides.

THE PHOSPHORUS CYCLE

All the elements that we have considered up to now have been oxidized or reduced by micro-organisms, but phosphorus exists in the environment almost entirely as forms of phosphate, usually orthophosphate; none of the other valency states are of importance. There have been reports of microbial oxidation and reduction of phosphorus e.g. to phosphine (PH_3), hydrogen phosphide (P_4H_2), phosphite (PO_3'') etc., but if they occur then it is on a very small scale. Another unusual feature of phosphorus is that it does not occur, except as dust traces, in the atmosphere since there is no common gaseous form.

Phosphorus is essential to all life and occurs in phospholipids, nucleic acids and ATP. It is not required in such large quantities as carbon and nitrogen, but it may be an important limiting nutrient. It is present in the soil at levels of 400 to 1200 mg/kg, though very little of this (less than 5%) may be available for it occurs as insoluble inorganic phosphates, particularly of calcium and iron, and in organic complexes such as inositol hexaphosphate. The concentration of organic phosphate decreases with increasing depth of soil.

Phosphorus is very much less common in aquatic environments (further reading 6, 8 and 10); there is about 0.01 to 0.07 mg/l though very much higher values are now recorded in some polluted waters (Chapter 7). Despite low concentrations the sea is one of the main phosphorus reserves, or perhaps it is best regarded as a final resting place of the phosphorus circulating in the environment for very little gets back out of the sea, except in guano and ultimately in geological movements which create new land. The phosphorus in aquatic environments may be as particulate matter, dissolved organic material, or inorganic phosphorus. The proportion of the different types varies with the particular body of water, the season of the year and the depth. The low natural levels mean that insufficient soluble phosphorus is present for the micro-organisms, particularly the planktonic algae, and they depend on the rapid mineralization or on the upwelling of lower layers of water (Chapter 7). The turnover time is a matter of minutes in surface waters during the summer or in the tropics, and at most usually only a few days. Despite these high rates the ability to use organic phosphorus direct, without waiting for mineralization, may be an advantage during algal blooms when phosphorus is particularly short. Bacteria growing epiphytically on planktonic algae may directly absorb exudates from the algae and the products of their autolysis. Some insoluble phosphorus does of course reach the sediments where iron and aluminium phosphates are common. The hydrogen sulphide produced under anaerobic conditions may release the phosphate again by precipitating ferrous sulphide.

The only input to the phosphorus cycle (Fig. 5.5) is from fertilizers, and a slow addition from the rocks, so the key processes in the cycle are the decay of organic matter and the dissolving of inorganic phosphorus. The decay of organic matter is carried out by a very wide range of saprophytes (Chapter 3) but the resulting release of phosphates is not the same sort of process as the mineralization of nitrogen where amino groups are converted to ammonium or nitrate. Phosphorus is already present as phosphate and is merely released by the breakdown of the carbon skeleton. The decay of nucleic acids and the phospholipids is quite rapid, but more complex organic compounds like phytic acid or polyphosphates stored by some bacteria and fungi may take much longer, particularly if they are adsorbed onto clay or colloidal humus particles. Some phosphorus is immobilized by the micro-organisms causing the decay: fungi contain 0.5 to 1.0% (by weight)

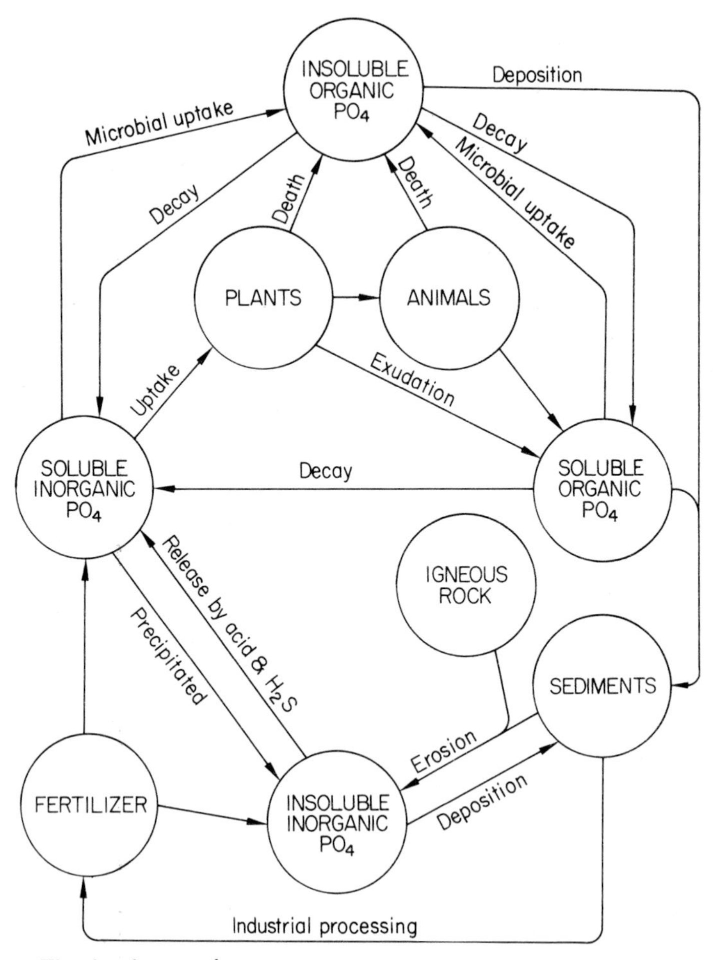

Figure 5.5. The phosphorus cycle.

and bacteria 1.5 to 2.5%. Higher plants contain 0.5 to 5.0%. There is therefore plenty of phosphorus in most decaying substrates and the concentration only needs to be greater than about 0.2% before some release occurs.

The metabolic acids released by micro-organisms can dissolve inorganic phosphorus in rocks; as many as 50% of heterotrophs isolated from soil have this ability, under labotatory conditions. The encouragement of this ability in soil micro-organisms has been investigated for agricultural purposes but the results are at present uncertain. It seems that most of the effects on plant growth can be attributed to other causes such as plant growth hormone production, similar to the free-living nitrogen fixing organisms (page 48). Micro-organisms living on or near the root (page 78) affect the uptake of phosphorus by the plant under some conditions: whether they increase or decrease uptake has been disputed. The investigation of this problem is fraught with difficulties of experimental technique and the results are frequently difficult to interpret. However it seems that at very low levels of phosphate plants may suffer a deficiency. With adequate phosphate micro-organisms are beneficial, particularly to young seedlings, which take up more phosphorus when grown with micro-organisms than without. There has been a lot of investigation of symbiotic associations between fungi and roots (mycorrhizal infections), particularly of forest trees, in relation to nutrient uptake in general and phosphorus uptake in particular. Here it is well established that uptake is increased in soils of moderate fertility; such is the beneficial effect of the correct mycorrhizal fungus that it is profitable to inoculate the young trees or their seed beds with the fungus, before planting new forests in soils which do not contain the correct species or strain. Mycorrhizae and other microbes growing on or near plant roots are considered again in Chapter 6.

In aquatic environments, particularly fresh water (further reading 6), phosphorus is normally a limiting nutrient and is important in the process of eutrophication (Chapter 7). Eutrophication can be a natural process, but it has come to be particularly associated with pollution by sewage and run-off from agricultural land fertilized with phosphorus and nitrogen.

IRON AND MANGANESE CONVERSIONS

There are two main groups of micro-organisms concerned in the conversions of iron and manganese. Firstly there are those, usually heterotrophs, releasing iron during the decay of organic matter and secondly there are autotrophic (chemosynthetic) bacteria which derive energy from the oxidation of the ferrous to ferric state (further reading 2).

Iron is present in organic matter in very small quantities, most importantly in cytochromes. During decay it may be released in a soluble form, immobilized or precipitated depending on the organisms concerned and the pH of the environment. Acid conditions, which can result from microbial metabolism, release iron from insoluble inorganic forms. There are some bacteria which accumulate iron or manganese from organic matter and deposit it as the hydroxide in their slime capsules. The taxonomy of these organisms is somewhat confused, but they are thought to be the main agents responsible for the deposition of iron and manganese in aerobic soils (e.g. *Pedomicrobium*) and in water (e.g. *Hyphomicrobium*). *Sphaerotilus* (and the closely related if not identical *Leptothrix*) and *Gallionella* are also heterotrophs which may deposit iron or manganese hydroxides. Their optimum pH is neutral or slightly acid and they are microaerophilic. There is some evidence that *Sphaerotilus* can grow chemosynthetically by oxidizing iron, and probably manganese, though it is normally found in waters rich in organic matter.

The truly chemosynthetic iron bacterium, *Thiobacillus ferrooxidans* (and *Ferrobacillus ferrooxidans* which is probably the same taxon) has a very low pH optimum (1.7 to 3.5), though it maintains an intracellular pH of about 5. It is a strict aerobe. The oxidation of ferrous to ferric yields only about 11.5 kcals, which is used to reduce (fix) carbon dioxide. It is thought that these bacteria are important in the geochemistry of iron and manganese, and micro-organisms have been implicated in, though not proved responsible for, the production of manganese nodules in the deep oceans. The use of micro-organisms in the extraction of low grade sulphide ores of copper, iron, nickel, cobalt and molybdenum is also possible on a commercial scale (further reading 11). *Thiobacillus ferrooxidans* (and probably *T. thioxidans*) oxidize the sulphide to sulphate, thus liberating the metals as soluble sulphates. Such reactions are responsible for the very acid drainage water from many mines which have a very restricted and specialized microflora including tolerant *Euglena mutabilis* strains, species of *Chlamydomonas*, the ciliate *Oxytricha* and the diatoms *Frustulia rhomboides* v. *saxonica* and *Ahnanthes microcephela*.

We have talked almost exclusively of the oxidation of iron, but there is some evidence that at low oxygen levels the ferric ion can be used as an electron sink, being reduced to the ferrous state. This will happen abiotically in environments with low redox potential, which may itself be caused by microbial activity. Thus in gley soils, which are usually waterlogged, the characteristic blue-green colour is caused by the ferrous ion.

The part played by micro-organisms in the geochemistry of other metal ions is much less clear but various changes in oxidation states have been reported (further reading 5). Such effects are most probably due to redox and pH changes induced by micro-organisms rather than direct metabolism of the ions.

THE SILICON CYCLE

The silicon cycle is mainly concerned with one group of algae, the diatoms (Bacillariophyta), which have shells of silicon dioxide. Radiolarians (Protozoa) also have silica shells and many organisms, including bacteria and fungi, and their lichen forms, can dissolve silicates from rocks, presumably by the production of acids. The amount of silicon immobilized by diatoms is small compared with that in the biosphere since igneous rocks, and their sedimentary derivatives, may be 90% silicon dioxide, the commonest forms of which are sand, and the clay minerals. Diatoms are comparatively rare in soils so we are concerned with the silicon cycle in water (further reading 6, 8, 10 and 11). The form of soluble silicon most important in water is the undissociated orthosilicate ($Si(OH)_4$) though there is also particulate matter, and colloidal aggregates of silicon tetroxide (SiO_4^{4-}) in more acid lakes though not in the sea. The concentration in water varies with the season of the year and with depth, and silicon may be limiting for diatom growth, particularly during the peak growth period in spring and early summer and during later blooms (Fig. 5.6). The concentration of silicon in the sea may have important secondary effects on other organisms that do not themselves require it as a nutrient, for it may be important in the long term regulation of the ocean pH. There is an equilibrium between aluminosilicate ($Al_2Si_2O_5OH$), calcium ions, potassium ions and potassium calcium aluminosilicate ($KCaAl_3Si_5O_{16}(H_2O)_6$). The involvement of calcium is then reflected in the levels of calcium carbonate and bicarbonate which in turn affects the carbonate/bicarbonate levels in general (Chapter 7).

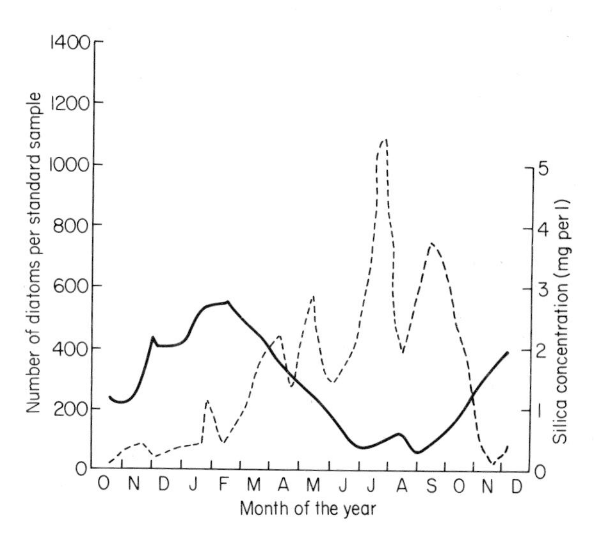

Figure 5.6. The variation in silica concentration and diatom numbers in Lake Windermere in the English Lake District. Silica concentration ———; diatom numbers – – – –. (From Round, F.E. (1960). Studies on the bottom-living algae in some lakes in the English Lake District. Part IV. The seasonal cycles of the Bacillariophyceae. *Journal of Ecology* **48**, 529–547.)

Silicon is taken up by diatoms, even from very dilute solution, and deposited directly in their frustules. The dissolution of the frustule after death is slow, much slower than pure particulate silicates, possibly because the frustule has organic coatings, though the cations adsorbed onto the surface are also important in retarding solution. A great quantity of diatom shells reach the sediments where they form diatomaceous earth which is used as an industrial abrasive.

This completes the discussion of the part played by micro-organisms in the cycling of nutrients within the biosphere. In conclusion it would be well to stress again that the many processes and reactions that have been discussed in separate sections or diagrams, or even in separate chapters, are interrelated in the real world. You should try to assemble in your mind's eye all the major components of the cycles into one dynamic system of interdependent equilibria. The stability of the equilibria, how they are controlled, what their weakest points are and therefore how close we are to upsetting the balance, are major problems in biology. The understanding of the micro-organisms involved is bound to play a very important part in the solution to these questions.

FURTHER READING

1 ALEXANDER M. (1961) *Soil Microbiology*. Wiley, New York and London. pp. 472.
Basic coverage of mineral cycles.
2 ARISTOVAKAYA T.V. & ZAVARZIN G.A. (1971) Biochemistry of iron in soil. In *Soil Biochemistry*. Vol. 2. Eds. A.D. McLaren and J. Skujins. Marcel Dekker, New York. pp. 527.
A review.
3 BROCK T.D. (1966) *Principles of Microbial Ecology*. Prentice-Hall, New Jersey. pp. 306.
Basic coverage of mineral cycles.
4 DOETSCH R.N. & COOK T.M. (1973) *Introduction to Bacteria and their Ecobiology*. Medical and Technical Publishing, Lancaster. pp. 371.
Basic coverage of mineral cycles, particularly sulphur metabolism.
5 EHRLICH H.L. (1971) Biogeochemistry of minor elements in soil. In *Soil Biochemistry*. Vol. 2. Eds. A.D. McLaren and J. Skujins. Marcel Dekker, New York, pp. 527.
A review.
6 HUTCHINSON G.E. (1957) *A Treatise on Limnology*. Vol. 1. Wiley, New York. pp. 1015.
Sulphur, phosphorus and silicon in lakes.
7 KELLOGG W.W., CADLE R.D., ALLEN E.R., LAZARUS A.L. & MARTELL E.A. (1972) The sulfur cycle. *Science* **175**, 587–596.
General review.
8 MARTIN D.F. (1968) *Marine Chemistry*, Vol. 2. Marcel Dekker, New York. pp. 451.
Particularly useful on silicon transformations.
9 MILLER J.D.A. (1971). *Microbial Aspects of Metallurgy*. Medical and Technical Publishing, Lancaster. pp. 220.
Iron and manganese, and corrosion.
10 WOOD E.J.F. (1965) *Marine Microbial Ecology*. Chapman and Hall, London. pp. 243.
Particularly for sulphur bacteria.
11 ZAJIC J.E. (1969) *Microbial Biogeochemistry*. Academic Press, New York and London. pp. 345.
Microbial effects on mineral deposits.

6 The structure and dynamics of microbial populations in soil

THE SOIL CHEMISTRY, STRUCTURE AND ENVIRONMENT

Soil is a very complex environment composed of three main phases, solid, liquid and gas, which are variously arranged, on both a macroscopic and a microscopic scale, to produce the hundreds of different soil types recognized in the world. Soil is formed by the physical, chemical and biological weathering of rocks to small particles, the mineral component, which with organic matter forms the solid phase. The chemical composition of the soil (Table 6.1) reflects that of the rock

Table 6.1 The chemical composition of soil. These are average figures from eight soils in various parts of the United States. (Modified from H.O. Buckmann and N.C. Brady, 1960. *The Nature and Properties of Soils*. 6th edition. Macmillan, New York. pp. 567.)

Constituents measured as:	% Composition	Comment
SiO_2	75.42	Sand particles and clay minerals
Al_2O_3	9.68	Clay minerals
Fe_2O_3	3.44	
K_2O	1.78	
Na_2O	0.96	
TiO_2	0.72	
MnO	0.11	
CaO	1.33	Limestones
MgO	0.85	
P_2O_5	0.11	
SO_3	0.10	
N	0.11	
Ignition	5.29	Loss of carbon on burning. (equivalent to organic matter content, unless the soil has a high $CaCO_3$ content which may break down to CaO and CO_2)

from which it was formed. Soils from humid or wetter regions would tend to have a higher proportion of organic matter and nitrogen than this example. The dominant mineral is silica, in sand, silt and clays, though it is possible for soils to have very high organic matter levels and virtually no silica (e.g. peats). Clay particles are composed of several different types of minerals derived by chemical weathering from the feldspars, micas etc. in igneous rocks and their sedimentary derivatives. Minerals occur in the soil as particles of various sizes, each of which may contain more than one mineral type; stones and gravel are above 2 mm diameter, sand 2 mm to 0.2 mm, fine sand 0.2 mm to 0.02 mm, silt 0.02 mm to 0.002 mm and clay below 0.002 mm. Soils can be named on the basis of their particle size distribution.

The organic matter may be more or less unaltered fragments of animals or plants, or may have been decomposed, eventually becoming humus (Chapter 3). The extracellular polysaccharides that many micro-organisms produce bind together the individual soil particles into crumbs; hyphae of fungi and actinomycetes are also claimed to play a part in this. The crumb structure is very important in soil drainage and therefore in aeration; a soil with a good, well developed crumb structure and/or a high sand content is freely draining and well aerated. In waterlogged, badly-managed agricultural, or eroded soils the crumb structure is usually poorly developed. Microbes generally live on particle surfaces or in the interconnecting spaces (pores) between the crumbs.

The soil water, in fact a weak solution of salts, is important for two reasons; firstly it is the solvent system in which plants take up mineral nutrients, and secondly the amount of soil water is inversely related to the amount of soil atmosphere, so that waterlogged soils are anaerobic. There are several different ways in which water is held in the soil; for example by impeded drainage, by adhesion and cohesion forces and by adsorption to soil particles. There is water in soil which is unavailable to plants and micro-organisms because of these physical forces and because the level of dissolved substances may prevent osmotic uptake.

The soil atmosphere occurs in those pore spaces not occupied by water, and is normally saturated with water vapour. The composition of the soil atmosphere depends on the biological activity and on the rates of gaseous diffusion and mass flow, which are themselves dependent on the solubilities of gases in water and on the pore size. There will be many spaces, in all but the most coarsely textured dry soils, that are not connected to the air, being cut off by continuous water films around particles. The supply of oxygen and the removal of carbon dioxide to and from these spaces is dependent on the slow diffusion through the water. In the soil atmosphere there is usually 10 to 100 times more carbon dioxide and rather less oxygen than in the air. Volatile organic materials e.g. methane, hydrogen sulphide, ammonia and hydrogen may also occur at greater concentrations.

Temperature fluctuations in sub-surface layers of soil are not as violent as those in air, though they still show annual cycles in the temperate and polar regions. Only in the latter are soils permanently frozen, or frozen to any great depth, and only in the dry tropical deserts does the temperature rise very high or the soil become dry to any great depth. The soil surface, particularly if there is no plant cover, is subject to more rapid and larger temperature changes. The surface temperature rises because of insolation, especially on dark coloured soils and rocks, and temperatures of more than 50°C occur in temperate regions while 70°C is known for the tropics or even up to 88°C for Death Valley, California. Similarly the surface temperature may drop very rapidly by radiation at night and the top few millimetres may be frozen while the bulk temperature is comparatively moderate.

The pH of the soil is largely determined by the chemical nature of the minerals and the organic matter present. Acid soils are most common in high rainfall areas with moorland or coniferous vegetation whereas alkaline conditions occur in dry grassland regions. The soil has a complex buffering system dependent upon the ion exchange capacities of the clays and colloidal organic matter, and the amount of insoluble carbonates (e.g. of calcium) or hydroxides (e.g. of aluminium) present; the buffering capacity of different soils therefore varies. Bulk soil pH may give a misleading impression of conditions because surfaces of clays may have a different, usually lower, pH by up to 1 or 2 units due to the adsorption of ions. Such variations over small distances may be very important to micro-organisms whose extracellular enzymes are also capable of being adsorbed and so may operate at a pH very different from that in the soil as a whole.

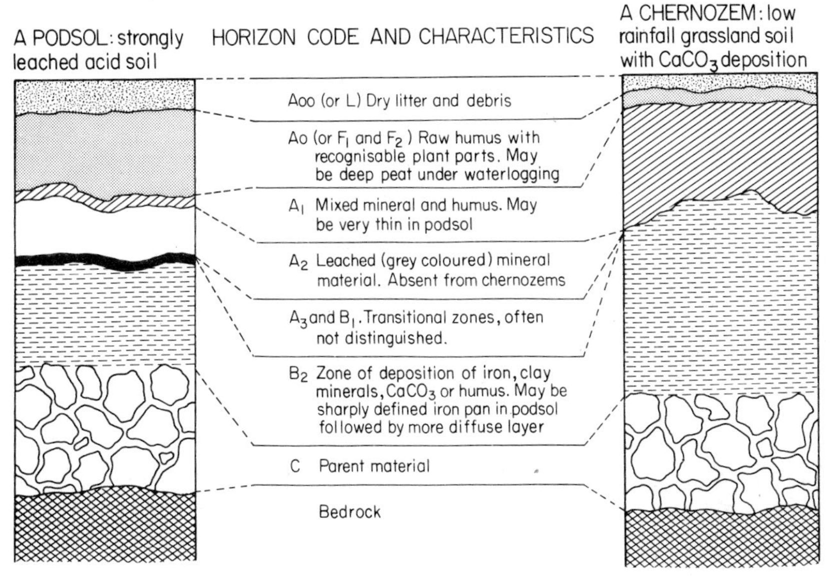

Figure 6.1. Diagrams of two soil types, a podsol which forms under conditions of acid leaching, and a chernozem which is characteristic of much drier conditions, often in grasslands. Some data on the organic matter content and the pH of such soils are presented in Table 6.3.

Some environmental factors vary with depth (e.g. temperature and aeration above). The chemical composition, particle size distribution and the concentration of organic matter also change with depth, sometimes markedly and often in a discontinuous way. Plant litter deposited on the surface decays and the downward movement of water through the soil causes leaching of soluble minerals, particularly if the litter is acid, and there is also displacement of particulate material. A sequence or more or less defined layers, called horizons, forms in the soil. In the simplest situation there is a surface layer of plant debris overlying minerals mixed with organic matter, with a mineral bedrock or subsoil. At the other extreme are soils with as many as eight or nine horizons in areas of high rainfall with acid litter. There is an internationally agreed code of letters for identifying the various horizons and soils are classified by the characteristics of their profiles (Fig. 6.1). There may be more variation between horizons in a single soil than there is between corresponding horizons of different soils.

THE DISTRIBUTION PATTERNS OF MICRO-ORGANISMS IN SOIL

Let us consider numbers and biomass first of all to gain some idea of the size of the population: one gram (dry weight) of agricultural soil contains several million bacteria, hundreds of thousands of fungal propagules and tens of thousands of protozoa and algae. The biomass of the fungi and bacteria may be nearly equal, and that of protozoa and algae an order of magnitude less (Table 6.2).

Table 6.2 The numbers and biomass of micro-organisms in the top 15 cm of agricultural soil. (Modified from R.C.W. Berkeley (1971) Microbiology of soil. In L.E. Hawker and A.H. Linton (Eds) *Micro-organisms, Function, Form and Environment.* Arnold, London. pp. 727.)

Organism	Number/g dry wt.	Biomass g/m^3
Bacteria	10^8	160
Actinomycetes	10^5 to 10^6	160
Fungi	10^5	200
Algae	10^4 to 10^5	32
Protozoa	10^4	38

These figures may vary greatly for other soil types; values for broadleaved woodland give much lower weights (at least an order of magnitude less) for all groups, with bacteria particularly low. Any of these figures for biomass are small compared with the standing higher-plant crop in an agricultural or forest situation.

The main determinant of microbial distribution on the small scale is the nature of the particles of which the soil is composed. This is partly a nutrient response, micro-organisms growing more on humus particles than on sand grains: in one study organic matter particles were only 15% of the soil, but had more than 50% of the bacteria associated with them. Some clay minerals have been shown to affect the growth rate of micro-organisms; for example montmorillonite increases the growth of bacteria but depresses the rate of linear growth of hyphae (further reading 12). There is also the purely physical effect of clays and organic colloids which adsorb bacteria, particularly in the presence of tri- or poly-valent cations. The net charges on the surfaces of the colloid and on the bacterium are important in determining the amount of adsorption. Fungal spores also have charged surfaces and so could be involved in adsorption. Nutrients are adsorbed, so concentrating them from the dilute soil solution, and pH effects at surfaces have already been mentioned (page 65). Surface phenomena are very important in soils, though often neglected by microbiologists.

Bacteria tend to grow as individuals or small microcolonies (often less than 10 cells) on the surfaces of soil particles (Fig. 6.2) and roots (Fig. 6.7). Fungi, and to a lesser extent actinomycetes, are different in that they can actively grow through the soil from a food base. While they may originate in an organic particle the hyphae can ramify through air spaces (Fig. 6.2) or regions of almost pure mineral. They do not usually grow as the discrete compact colonies seen in pure culture. Instead a hypha grows, comes upon some utilizable organic matter,

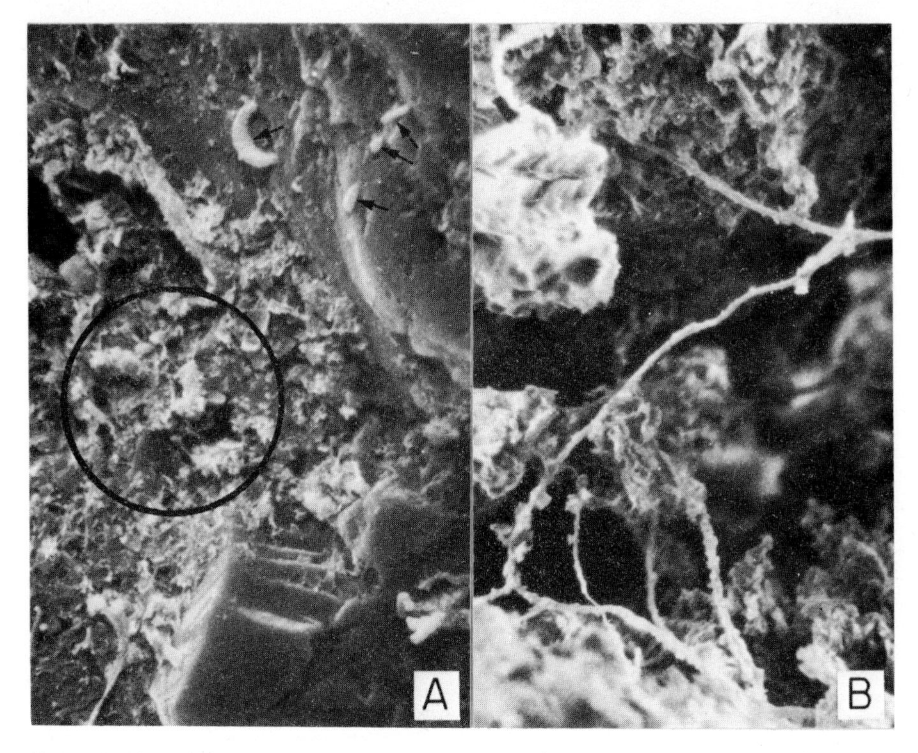

Figure 6.2. Scanning electron micrographs of bacteria (arrows) on soil particles and hyphae growing through the soil pores. Micro-organisms are often coated with mucilage and clay cutans which may make them very difficult to see (e.g. some of the bacteria circles in A). (A) × 2,500. (B) × 1,400.

covers the surface by branching and as the substrate is exhausted hyphae grow off again: the original hypha may well have died and been lysed. There is therefore a discontinuous advancing edge, often with nothing alive behind it. Most of the microbial growth in the soil, as opposed to survival, is restricted to a few scattered microhabitats. Sporulation of fungi occurs on the soil or litter surface or in the larger pore spaces. The distribution of soil algae and protozoa depends on the morphological type. The unicellular forms occur singly in the water films, particularly if motile like many of the protozoa, the flagellate algae and the diatoms. Filamentous algae may occur in almost single-species, discrete colonies which can reach macroscopic size and be seen on the soil surface with the naked eye.

Since the majority, if not all, microbes live in water films they are primarily affected by those aspects of soil structure which determine the water conditions. Even in dry soils in deserts where there is apparently no free water, the relative humidity in the soil atmosphere is usually greater than that of the surrounding air, and spores, cysts and other resistant structures may survive for long periods. The motile protozoa and algae, particularly flagellates, some specialized ciliates (e.g. *Disematostoma* spp., *Spiretta*, *Spirofilopsis*, etc.), *Euglena* and members of the Volvocales respond very quickly to free water and may become numerous in or on even temporarily wet soil or in rain puddles that persist for a few days.

The responses to the level of soil aeration have been mentioned in the preceeding chapters on the nutrient cycles. Fungi and most protozoa are aerobes while some bacteria are obligate or facultative anaerobes. Though the bulk of soil is only rarely anaerobic or microaerophilic, there are steep oxygen and carbon dioxide gradients in and around the microcolonies, and anaerobic or microaerophilic microenvironments occur even near the soil surface where particulate and dissolved organic matter is being rapidly oxidized. A simple relationship between the depth in the soil and the proportions of aerobic and anaerobic organisms does not therefore exist (Table 6.3, page 70). In culture low levels of oxygen and/or high levels of carbon dioxide have morphogenic effects on fungi,

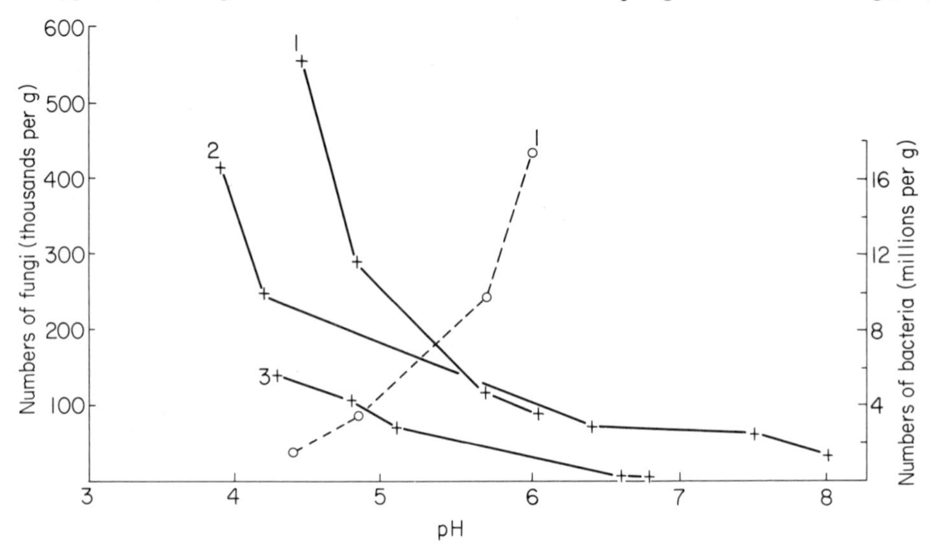

Figure 6.3. The relationship between soil pH, the numbers of bacteria and the number of fungal propagules (measured by dilution plates). +——+ fungi: ○----○ bacteria. (Plotted from data of (1) A. Starc (1941). Mikrobiologische Untersuchungen einiger podsoliger Böden Kroatiens. *Archiv für Mikrobiologie* **12**, 329–352. (2) J.H. Warcup (1951). The ecology of soil fungi. *Transactions of the British Mycological Society* **34**, 376–399. (3) D.M. Webley, D.J. Eastwood and C.H. Gimingham (1952). Development of a soil microflora in relation to plant succession on sand dunes, including the 'rhizosphere' flora associated with the colonizing species. *Journal of Ecology* **40**, 168–178.)

such as the change from hyphal to yeast form in some *Mucor* species. The form observed on agar plates may thus be very different from that in the soil, and conversely direct examination of the soil may reveal unidentifiable micro-organisms which are in fact just different growth forms of common species.

The role of pH in the distribution of soil microbes is very complex because it has such wide ranging effects on leaching, nutrient availability, adsorption, extracellular enzymes etc.; however some generalizations can be made. Fungi have rather lower pH optima for growth than most bacteria and this is reflected in their occurrence: in most acid soils (below about pH 5) fungi tend to be more important than bacteria (Fig. 6.3). In some acid soils it is the proportion of fungi which increases, compared with alkaline soils, though the absolute numbers may decrease. There are exceptions, of course, such as the extremely acid tolerant sulphur oxidizing bacteria (*Thiobacillus*), and the pH effect may be overshadowed by the effect of organic matter or plant roots. Cyanophyta are rare in more acid soils; they are most common in slightly alkaline conditions though some show extreme tolerance in alkaline deserts. Algae are more variable in their response to acidity; Chlorophyta are mostly tolerant to a wide range, though some are particularly acid tolerant (e.g. desmids, *Cylindrocystis* and *Maesotaenium*). Bacillariophyta (diatoms) tend to be more frequent on neutral or slightly alkaline soils (*Epithemia zebra* and *Rhopalodia* spp.) though there are notable exceptions which are characteristic of acid soils (e.g. *Calonies fasciata*). Amongst protozoa, rhizopods (amoebae, especially testate amoebae) are characteristic of acidic sites while Ciliophora (ciliates) are normally found in neutral and alkaline soils.

Temperature is one of the main controlling factors for microbial activity, though there are psychrophilic and thermophilic populations associated with arctic and tropical soils and with specialized habitats such as compost heaps, where microbial activity itself causes the rise in temperature. Many of the algae and Cyanophyta can withstand the extreme fluctuations of temperature at the soil surface and the associated problems of water loss. Some algae will even grow on frozen soil or snow. Even if soil temperatures go outside the normal growth range many micro-organisms can survive as spores or cysts.

The last major variable in the soil environment is the nutrient status. For algae and other autotrophs this means the levels of minerals. Organic matter can be inhibitory to algae but some (e.g. *Prasiola crispa*) are particularly associated with habitats rich in organic matter such as guano deposits. The abundance of most protozoa is dependent on bacterial numbers, though the testate amoebae may be directly dependent on organic matter. The critical factor for all heterotrophic bacteria and fungi is the quantity of organic matter, and to some extent its quality (Chapter 3). Most organic matter is insoluble and unevenly distributed in the soil, and the numbers of bacteria and fungi are often correlated with this distribution (Table 6.3).

Part of the soil flora is in a relatively continuous, if low, state of activity utilizing the more intractable, insoluble materials such as lignin and humus. This group (the autochthonous flora) includes ascomycetes, basidiomycetes, actinomycetes and some bacteria. The other heterotrophs which utilize simpler, often soluble, substrates are normally dormant and only become important when fresh organic matter is added to the soil. The activity of the latter flora (the zymogenous flora) therefore varies greatly over time and in space, though they are always present in a dormant state.

Many of the environmental factors just considered act on the soil micro-organisms according to seasonal rhythms (e.g. temperature fluctuations and organic matter supplies in temperate regions). The microbial activity therefore follows a seasonal pattern, though it may lag behind the causative factor. Climatic factors are generally more important in seasonal variations than organic matter supply.

Let us now consider examples of estimates of numbers in different soils to see if environmental factors can explain some of the observed variation in abundance. The most developed soil type, in terms of the number of horizons, is a podsol and an example of the microbial population is given in Table 6.3. The pH is not

Table 6.3 The number of various groups of micro-organisms, measured by dilution plates, in relation to the horizons, organic matter concentration, pH and soil types. Soil profile 1 is a podsol in a mixed broadleaved-conifer forest; soil profile 2 is of a chernozem with grass cover; soil profile 3 is a peat with *Sphagnum* and *Polytrichum*. (Selected from M.I. Timonin (1935) The micro-organisms in profiles of certain virgin soils in Manitoba. *Canadian Journal of Research* **13C**, 32–46.)

Soil	Horizon and Depth (ins.)	Moisture %	pH in water	Organic matter %	Aerobic bacteria 10^6/g D.W.	Actino-mycetes 10^6/g D.W.	Anaerobic bacteria 10^6/g D.W.	Fungi 10^3/g D.W.	Algae 10^3/g D.W.	Protozoa 10^3/g D.W.
1	A_0 0–3.5	48.7	6.55	64.17	28.30	5.70	0.10	242.50	1.00	0.10
Podsol	A_1 3.5–5	23.1	5.51	25.00	4.50	3.50	1.00	20.00	0.10	0.02
	A_2 5–8	13.6	5.50	4.47	1.55	0.95	0.01	1.63	0	0
	B 8–23	19.1	6.37	2.60	3.30	0.90	0.10	10.63	0	0
	C 23–37	18.7	7.81	0.95	1.14	0.12	0.01	1.47	0	0
2	A_1 0–2	82.0	7.46	22.25	19.00	3.25	1.00	60.13	1.00	0.10
Chernozem	A_2 2–6	24.3	8.08	8.64	16.50	3.00	1.00	6.00	0.50	0.02
	B_1 6–16	31.7	8.09	2.45	16.73	0.65	1.00	2.50	0	0
	B_2 16–27	31.7	8.25	1.27	2.51	0.15	0.001	0.20	0	0
	C 27–44	18.8	8.27	0.31	0.28	0	0.001	0.04	0	0
3	1. 0–18	500.7	4.78	90.24	2.90	1.15	0.10	372.50	10.00	10.00
Peat	2. 30–37	620.2	5.43	68.00	2.74	0.01	0.10	3.10	0.50	0
	3. 70–76	750.0	4.39	88.91	0.06	0	0.10	0.88	0	0

particularly low for a podsol, but the fungi are nevertheless important, particularly in the litter layer where raw organic matter is present at high levels. The anaerobic bacteria are most numerous in the layers near the surface. At depth, where there are low organic matter levels, and in the leached A_2 horizon, there are comparatively low numbers of all heterotrophs, including anaerobes. The rise in numbers in the B horizon, e.g. the ten-fold increase in fungi, is a common feature of podsols and generally reflects deposition of nutrients and the higher pH in this horizon. The very high figure for aerobic bacteria in the A_0 is probably spurious (Timonin comments on it in his paper) and is an illustration of the variability in dilution plates which can only be overcome by taking many replicates.

Chernozem soils (Table 6.3) are less acid than podsols and are usually richer in available nutrients. The bacterial numbers in equivalent horizons are therefore higher and the fungi, even though their numbers are similar in both soils, are a lower proportion of the population.

Peat has a large proportion of fungi and the bacterial numbers, including the actinomycetes and the anaerobes, are very low, probably because of the low pH. There are relatively high numbers of algae and protozoa in the surface horizons and this may be linked with the high water content. The protozoa are likely to be mostly testate amoebae, though there is no information on this.

These comparisons bring out three general points. Firstly, they show the difficulty of comparing soils even if, as in this case, the horizons are known and the figures are all from one worker using one set of methods. Secondly, the numbers of heterotrophs are closely correlated with the organic matter levels except under extreme conditions such as the peat which has a low pH: this is a good example of the limitation of a population by the least favourable factor in the environment (Chapter 2, page 14). Thirdly, protozoa and algae are unimportant in these soils, even at their maximum development in the peat. These data provide no information on the chemotrophic bacteria but their numbers would be low in comparison with the heterotrophs.

TYPES OF MICRO-ORGANISMS IN THE SOIL

We have been discussing 'bacteria', 'fungi', 'algae', and 'protozoa' but we must now consider what families, genera or species are most commonly encountered. Almost any micro-organism can be found in soil if the search is sufficiently diligent: for example, pathogens may arrive in or on the soil in a dead animal, but they do not generally survive and are not part of the soil population. Here we are concerned only with those micro-organisms that normally pass at least some part of their life cycle in the soil.

Soil bacteria are often not well defined by many of the common criteria that medical bacteriologists have developed as the basis for classification. Thus there are many pleomorphic forms (they change their shape according to nutrition and other environmental factors) and some also show a variable reaction to Gram's staining technique: these are usually grouped in the genus *Arthrobacter* in the Corynebacteriaceae. Recent work on the numerical taxonomy of the Corynebacteriaceae has suggested that *Arthrobacter* be removed from this family.

Coryneforms usually comprise at least half of the colonies on aerobic dilution plates with unselective media. A further 25% of the population will be spore formers (*Bacillus*). The anaerobes have been very little studied except in general lists (Table 6.3) where they are 5% or so of the population. Actinomycetes, particularly *Streptomyces*, are apparently common and are readily isolated from aerobic, neutral or slightly alkaline soils where they may constitute up to 20 or 30% of the bacterial numbers. The genera just discussed account for almost all the colonies regularly isolated and yet there are lots of bacteria not mentioned e.g. *Pseudomonas*, *Nitrobacter*, *Nitrosomonas*, *Azotobacter* and *Rhizobium*, which are thought to be, and probably are, very important in the soil nutrient cycles. Altogether these are less than 10% of the colonies isolated, and most of this 10% are *Pseudomonas*.

The vast majority of fungi isolated are those forming large numbers of spores, particularly Mucorales (*Mucor*, *Mortierella* and *Rhizopus*) and deuteromycetes (*Penicillium*, *Aspergillus*, *Cladosporium*, *Fusarium*, *Alternaria* and *Botrytis*). However direct examination shows many dark coloured hyphae, possibly deuteromycetes or more likely sterile mycelia. Though infrequently isolated, basidiomycetes are nevertheless important in the breakdown of organic matter and occur widely, as evidenced by the large fruit bodies of the Agaricales and Aphyllophorales that appear most years. Basidiomycetes also form mycorrhizal associations (see below) with trees and this accounts in some measure for their distribution; for example *Russula* and *Lactarius* are often found in beech and pine woods.

The genera of soil protozoa are similar to the aquatic ones, though the smaller species tend to be more common, or even smaller individuals of a given species. There are only a few that are found exclusively in soil. Species common to most soils include *Heteromita globosa*, *Oikomonas termo* and *Cercomonas* spp. (all flagellates): *Colpoda cucullus* and *C. steinii* (ciliates): *Naegleria gruberi*, *Acanthamoeba* spp. and *Hartmanella hyalina* (amoebae) and a few more, such as the testaceous rhizopods, which are almost entirely restricted to acid soils (further reading 4, 8 and 11). The protozoa are most frequent in the surface horizons (Table 6.3) and are almost always cyst forming types. Many of the slime moulds (Myxomycetes) and the cellular slime moulds (Acrasiomycetes) occur in the soil and their amoeboid stages are indistinguishable from simple free-living amoebae. The importance of slime moulds has been little studied, but even if they are not present in large numbers (hundreds, or occasionally thousands/g), some of the species are very widely distributed. Much work remains to be carried out on the ecology of protozoa since they vie with the algae for being the least studied of soil micro-organisms, largely because of methodological problems.

Actively growing algae (further reading 4 and 10) are confined to the soil surface or at best the top few centimetres. There is therefore not much scope for vertical distribution patterns though there are reports that even in the top 1 cm there is a microstratification, with the Xanthophyta being deepest in the soil. All algae are washed down the profile and some may live at depth heterotrophically (e.g. species of the Bacillariophyta and Chlorophyta) though the importance of this facultative heterotrophy is doubted by some and is certainly more a matter of survival than active growth: it is not a major path of carbon breakdown. A large number of species have been reported from soils, but only comparatively few are common. There is a distinct soil population; *Chlorococcum humicola*, and some *Oedogonium* and *Vaucheria* spp. are unknown from water. Soil forms are usually smaller than the related aquatic species or strains and most are cosmopolitan (e.g. *Hantzschia amphioxus* and species of *Nostoc, Chlamydomonas, Stichococcus, Zygogonium, Hormidium*) though Chlorophyta are more usually dominant in temperate conditions and Cyanophyta become more common, though not necessarily dominant, in the tropics.

Viruses have not been mentioned before, but bacterial viruses (bacteriophages) occur in soil, widely spread though not apparently in large numbers. They are not thought to have any significant effect on heterotrophic bacterial numbers, though there is little information. There are 'phages that attack *Rhizobium* and these have been studied in relation to nodulation and nitrogen fixation. Unfortunately the rhizobia that are 'phage resistant are often those that are not efficient at nitrogen fixation. Viruses pathogenic to higher plants also occur in the soil; some survive but a short time and are presumably destroyed by heterotrophs. Others such as tobacco mosaic virus are known to survive for years on infected plant remains. Adsorption to organic matter and clay particles seems to be important in long term survival. There are root infecting viruses which are actively transmitted through the soil by their vectors, notably nematodes. Some, such as tobacco necrosis virus and lettuce big vein virus, are transmitted by the fungus *Olpidium brassicae* which is itself a root pathogen. Apart from being vectors fungi are themselves parasitized by viruses and they have been reported from *Penicillium* spp., the cultivated mushroom (*Agaricus bisporus*), and from *Gaeumannomyces graminis* where the viruses have been claimed to be partly responsible for the loss of virulence in some strains of this pathogen of cereals and grasses. Soil algae and protozoa are also reported to be attacked by viruses but their importance is unknown. Rather surprisingly there is also very little information published on the occurrence or survival of animal viruses in soil. They probably do not persist for very long under temperate or tropical conditions, a matter of a few days. Russian work on foot and mouth disease has shown this to be true in the Siberian summer, particularly if the virus is exposed to sunlight. During the winter, however, the survival rate is much better and it is possible for infected soil to remain so from autumn, through the winter and into spring and early summer. Various arthropods are known to be vectors or reservoirs of mammalian viruses which exist in soils.

SUCCESSION, COMPETITION AND PREDATION IN SOIL POPULATIONS

These subjects were discussed in general terms in Chapter 2. Variations in the soil population with physical, seasonal and environmental factors have been mentioned, but superimposed on these fluctuations are progressive changes in the communities which are part of seres or other successions.

Micro-organisms are directly involved in the development of structured soils from bare rock surfaces or eroded subsoils, desert surfaces etc., habitats characterized by very low levels of soluble mineral nutrients, the almost complete absence of organic matter and extremes of temperature and moisture. The initial colonizers are therefore a few tolerant algae and Cyanophyta, particularly those with mucilaginous walls or slime capsules. The ability to fix nitrogen is also a great advantage. Typical organisms are *Porphyrosiphon notarisii, P. cinnamomeus, Zygogonium ericetorum, Gloeocapsa* spp., *Gloeocystis, Palmogloea protruberans, Nostoc ellipsosporum, N. muscorum, Schizothrix calcicola* and *Scytonema hoffmanii* (further reading 10). Organic matter accumulates from the organisms and wind-blown dust and the surface is stabilized, particularly by those species with extracellular mucilage. Infiltration and water retention of primitive soils is improved and, when moist, the algal crust is a medium in which seeds of higher plants will germinate. On bare rock surfaces the algae, fungi and bacteria may dissolve minerals e.g. silicates, and so hasten the chemical changes. By growing within fissures and cracks they can also accentuate the physical weathering due to expansion and contraction of water containing tissues during freezing-thawing and wetting-drying cycles. During life the algae supply soluble organic and

Table 6.4 The accumulation of minerals by lichens growing on rocks. (From J.K. Syres and I.K. Iskander (1973) In V. Ahmadjian and M.E. Hale (Eds) *The Lichens*. Academic Press, New York and London. pp. 225–248.)

Lichen species	N %	P μg/g	K μg/g	Fe μg/g
Verrucaria sphinctrina	0.72	486	1140	810
Caloplaca citrina	2.42	2000	3990	5020
Aspicilia calcarea	1.31	695	1530	2680
Physcia caesia	1.90	1800	3250	4140
Xanthoria parietina	1.98	1320	3950	3210
Limestone rock	Trace	48	120	230

mineral nutrients by leaching or exudation and their dead remains are also left in the developing soil. These may be used by saprophytic, or occasionally parasitic, bacteria and fungi. There are also more organized mutualistic relationships between algae and fungi which are called lichens. The two components of the lichen form a very stable association with different macroscopic appearance from either symbiont. In simple terms the fungus obtains carbohydrates from the alga and the latter may gain a measure of protection from desiccation since the hyphae frequently have gelatinous walls. Lichens have been claimed to be important in increasing the rate of soil formation from bare rock, though their growth rate is so slow that their effect must be small compared with that of abiotic agents. They may accelerate physical destruction of the rock by shrinkage and expansion of the thallus, and lichens also produce a wide range of substances which may chemically degrade the rock including carbon dioxide (acting as H_2CO_3), various organic acids and chelating agents. Lichens may accumulate minerals and nitrogen (Table 6.4) which are eventually released to the primitive soil when the lichen thallus is decayed.

As the amount of algae and lichens increases there is a rise in the numbers of fungi and bacteria. Protozoa probably come into the population in significant numbers quite late in the sequence, though one genus, *Leptopharynx*, has been associated with comparatively early stages of algal mats on eroded soil. Soil arthropods and their faeces (see page 77) are also important in soil formation and fertility, though they often depend on specialized protozoa in their gut (e.g. Collembola, the springtails) which utilize the ingested algal cells.

All these mechanisms may therefore lead to the build up of a primitive, and very thin, soil which has considerably more nutrients, particularly organic materials, than the parent rock or mineral fragments. It is, however, very unstable and the sere may be set back to zero many times before the community has sufficient diversity of micro-organisms and nutrients to support the growth of higher plants whose roots will help further to stabilize the developing soil.

The sere may be shortened if organic matter can be added from outside the the system. In recent studies of the newly formed volcanic island of Surtsey, organic matter was added to the littoral zone by the sea and fungi and bacteria rapidly became established in the sand and volcanic ash, closely followed by higher plants. Algae were important in rock and ash colonization but lichens only arrived some eight years after the start and were not important in the succession.

Within the long term sere from primitive to mature soil there are many short term changes in the microbial populations which reflect the colonization of newly arrived pieces of organic matter. These are termed substrate successions. In soil the organic matter is usually plant litter (leaves, twigs, tree trunks, roots etc.) though animal bodies, particularly the regular supply from the many soil arthopods, are also important. All these materials arrive with a resident microflora on or in them, though this imported population does not usually survive the extreme competition that occurs in the soil for these substrates. The long term successions discussed above have environmental constraints, such as nutrient deficiency, and only slight competition at the beginning but the complexity increases during development. They are basically different from substrate succession on litter where competition decreases as the organic matter is used.

Though the details of the genera or species will vary, the substrate succession usually starts off with the breaking of dormancy of the spores and other resistant structures in the volume of soil affected by the organic matter. There is a great increase in the activity of the zymogenous fungi (the so-called sugar fungi) such as *Mucor, Rhizopus, Trichoderma, Penicillium* and *Aspergillus*, and among the bacteria the pseudomonads and bacilli are probably the most important. These micro-organisms have a very high competitive saprophytic ability and thus possess many of the characteristics of effective competitors (Chapter 2, page 15). As the easily utilizable materials are consumed, and the pH drops due to the immobilization of cations, the population slowly changes to ascomycetes, some actinomycetes, coryneforms and *Bacillus* spp. This group may also include *Cytophaga*, myxobacteria and *Streptomyces* which can lyse and degrade the hyphae which were part of the initial invasion of the substrate. Most bacteria are also used as food by protozoa which increase in numbers during the initial stages of the substrate succession. The final stage in the succession is the invasion of the now much depleted substrate by the autochthonous organisms. In general the species diversity of both bacteria and fungi decreases as the material is decayed.

There may be commensalism rather than competition, particularly in the later stages of the succession. Thus many basidiomycetes have vitamin requirements (e.g. for thiamin and biotin) which in culture can be supplied by other organisms that are, or have been, growing on the same substrate. Mutualistic relationships also occur: different microbes, or different strains of the same species, may possess complementary enzymes. A single organism could rapidly exhaust its available substrate if it was complexed with other unavailable material, but groups of compatible organisms may between them have the whole range of enzymes necessary to effect the complete breakdown of the mixed substrate.

An example of a succession of fungi is given in Figure 6.4. There are very few

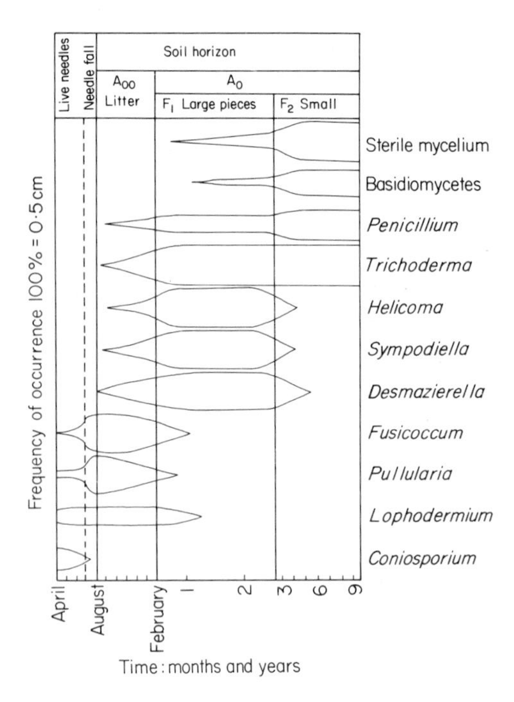

Figure 6.4. The sequence of colonization of pine needles: note that the time scale is not linear. (From W. Bryce Kendrick and A. Burges (1962). Biological aspects of the decay of *Pinus sylvestris* leaf litter. *Nova Hedwigia* **4**, 313–344.)

detailed data on bacteria, but they are present on the plant material when it dies and they probably form a part of the first population. In wood in contact with the ground, and in standing timber, bacteria are among the first colonizers and may be synergistic or mutualistic to later arrivals. They increase the permeability of the wood by destroying pit membranes and there is the possibility that some of them fix nitrogen, a nutrient which is in particularly short supply in decaying wood.

These successions are rarely uniform invasions over the whole substrate, particularly if it is a large piece. Initial colonization and nutrient depletion may be rapid in one spot where a leaf surface, for example, is broken. At other points the cuticle may be intact and form at least a temporary barrier to invasion. A mosaic of colonization therefore develops with different parts of the substrate in different stages in the sequence of activity by the various colonizers. There is a little evidence that successions also involve chemotrophic bacteria such as are concerned with the nitrogen and sulphur cycles. Thus when ammonium or hydrogen sulphide have been produced from the heterotrophic breakdown it is possible for *Nitrosomonas*, *Nitrobacter* and the sulphur oxidizers to become active at the particular sites or more generally on the substrate.

Different sorts of organic matter have different populations of micro-organisms. This is hardly surprising for obviously different substances such as animal versus plant material, but it is true also for leaf litter derived from different tree species. The specificity of some fungi is probably caused by differences in the chemistry of different plant tissues and in the physical structure, which affects aeration for example. Conversely there are some fungi which seem to have a very wide tolerance and occur on many different substrates (further reading 13).

Some mention must be made of organisms apart from microbes, particularly the soil arthropods and earthworms. These chew up the litter and so break it into small pieces in which the surface area available for microbial colonization has been greatly increased. Passage through the gut removes comparatively little of the original organic matter, though there may be bacteria and protozoa in the gut which assist in the breakdown of intractable substances. The addition of mucilage, and probably uric acid, to the comminuted organic matter increases the rate of microbial colonization of the voided faeces. The microbial activity may be several times greater than that which would be expected on similar undigested organic matter. Faecal pellets are probably most important in increasing the rate of turnover of organic matter in the soil. Coprophagous animals may derive nutrients from the microbes that have colonized the faeces, voiding the organic matter again to be recolonized. Soil animals also disturb the soil, moving organic matter about and exposing new surfaces; they carry micro-organisms on their bodies, so inoculating new material.

The soil population fluctuates in relation to a variety of factors, especially climatic and successional ones. Nevertheless it is an apparent paradox that the overall soil population is remarkably stable, though composed of a mosaic of local interactions in dynamic equilibrium with the environment and neighbouring microbial communities. The interactions between the micro-organisms can be considered in terms of food webs (Chapter 2) for, though many microbe-microbe interactions owe much to antagonism and competition, there are predator-prey relationships as well (e.g. protozoa ingest other micro-organisms and the fungi *Arthrobotrys* and *Dactylaria* trap nematodes). Some of these relationships in successions and food webs are summarized in Figures 2.2 (page 11) and 6.5. Incidentally, though these concepts of competition and predator-prey relationships are useful, they have been mostly studied in laboratory conditions because of the awful complexity of the natural situation. There is in fact very little information on such activity in the soil itself, or even about the effect of environmental factors such as pH, Eh etc. on the balance between different organisms or groups of organisms on a microhabitat scale.

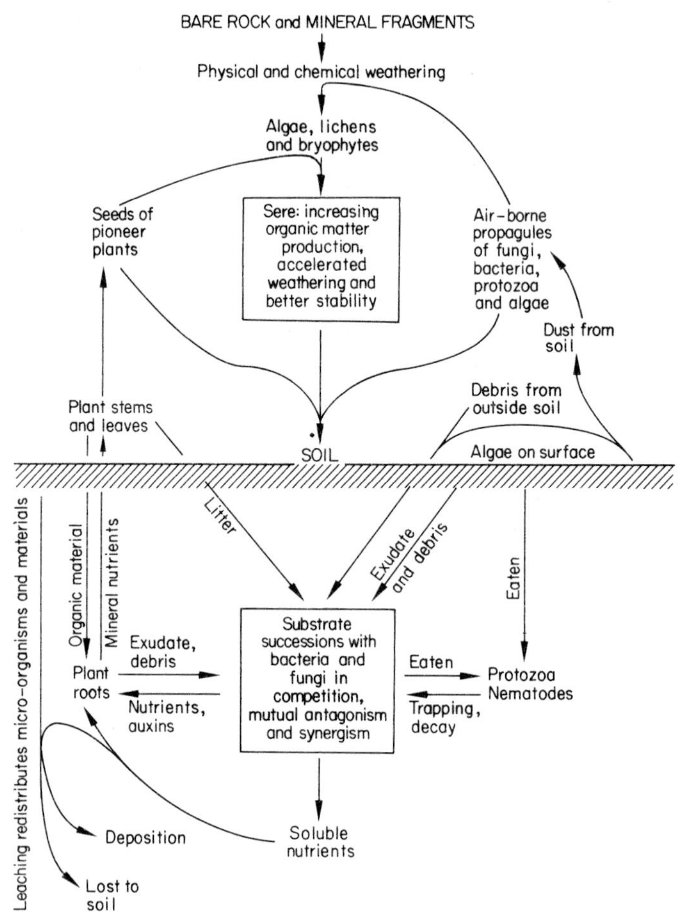

Figure 6.5. Summary of the interrelationships between organisms in the processes of soil formation and development.

THE RHIZOSPHERE

The effect of the plant on micro-organisms

Figure 6.5 has brought the plant roots into the soil picture and we must now consider these in more detail for they have a profound effect on micro-organisms. The volume of soil affected by plant roots is called the rhizosphere (further reading 3 and 14) and its limit in one direction is the root surface, the rhizoplane. This idea is fairly easy to envisage theoretically but the practical limits of the rhizosphere are more difficult to define. In experiments the outer limit is often determined very crudely: at worst the plant is dug up, the loose soil shaken off and that which remains on the roots is the rhizosphere. This may be refined in special experimental situations by sampling at known distances from the root, but this is difficult because of the small distances involved. Serial washings of the roots give some distinction between parts of the rhizosphere, the longer or more vigorous the washing the nearer to the rhizoplane is the soil which is removed. Vigorous shaking with glass beads for 10 to 15 minutes is necessary to remove most of the rhizoplane flora. In some situations, such as pot-bound plants or under permanent grassland, the roots may be so dense that all the soil is effectively rhizosphere. The rhizoplane is apparently easier to define, but even here there are some problems because micro-organisms can invade the epidermal and cortical cells, particularly if they are damaged, so that rhizoplane organisms can be found within the root (Fig. 6.7B, page 84).

The root influences the soil in several ways: the carbon dioxide levels are raised by root respiration and the oxygen, water and mineral concentrations are reduced. The major cause of the rhizosphere effect is however considered to be the organic nutrients added to the soil by the root. There are cells from the root cap, and wall material from the cortex when secondary thickening of the root occurs. The soluble material is a complex chemical mixture and exudation occurs mostly in the zones of elongation and root hair growth. Estimates of the total amount of material added to the soil vary greatly, but it is probably 1–2% of the carbon translocated to the root and only 20% of this is soluble.

The list of soluble compounds from sterile grown plants is extensive (further reading 3), though most of the information is from young seedlings and there is very little known about mature plants. Simple sugars such as glucose and fructose are almost universal and pentoses, di-, tri- and oligo-saccharides are all reported: it is probable that high molecular weight polysaccharides are also important. All the common amino acids occur but alanine, serine, leucine, valine, glutamic and aspartic acids, glutamine and asparagine are produced in the greatest quantities by the most plants. Organic acids are found, but not in large quantities. This covers the major part of the exudate but there are many substances present at very low levels which may nevertheless be important in the nutrition of micro-organisms growing in the rhizosphere: these include vitamins, particularly thiamin and biotin, nucleotides, flavones, auxins and assorted, chemically ill-defined, stimulators or inhibitors of particular microbes. Foliar sprays of some nutrients, herbicides, nematocides and antibiotics may be translocated to, and exuded by, roots.

The situation under non-sterile conditions becomes more complicated, for micro-organisms use and modify the compounds and exude their own soluble substances, some of which may affect the root and cause morphological and permeability changes (see below). There is very little information on the chemistry of the exudates under normal soil conditions but it is possible, indeed probable, that they are different from those released under the sterile conditions described above. Recent work also indicates that the root may produce a number of volatile substances (e.g. ethanol, acetaldehyde, isobutanol, alkyl sulphides) which influence the microflora. The species or strain of plant, the age, physiological condition and environmental factors can all affect the quality and quantity of root exudates.

The release of all this varied organic matter to the soil greatly increases the overall numbers of micro-organisms, mostly those from the soil, though some seed-borne organisms may be able to establish themselves. The soil population does not however react uniformly. Some groups are apparently unaffected, such as the autotrophic algae which are anyway not important below the soil surface.

Amongst bacteria, *Azotobacter* is one of the few that is usually unaffected (though see example in Table 6.5). A decrease in numbers in the rhizosphere is shown by Gram positive cocci, pleomorphic forms and sometimes rods including *Bacillus*. However the commonest response in bacteria is an increase in numbers, compared with the soil in general, and this has been shown for Gram negative non-spore forming types (*Agrobacterium, Corynebacterium, Pseudomonas, Rhizobium*) and many of the nitrogen cycle bacteria. The fungi also increase, particularly the Mucorales in the exudation zone, and cellulose decomposers where cells are sloughed off; there are different populations at different distances from the tip

Table 6.5 The changes in numbers of various groups of micro-organisms in the rhizosphere of wheat compared with soil in which no roots were growing. The R:S ratio is the ratio of numbers of organisms in the rhizosphere to those in unrooted soil. (From J.W. Rouatt, H. Katznelson and T.M.B. Payne (1960) Statistical evaluation of the rhizosphere effect. *Proceedings of the Soil Science Society of America* **24**, 271–273; by permission of the Soil Science Society of America.)

	Numbers per g dry weight		
Micro-organisms	Rhizosphere soil	Control soil	R:S ratio
MAJOR TAXONOMIC GROUPS			
Bacteria	1200×10^6	50×10^6 †**	240.0
Actinomycetes	46×10^6	7×10^6 **	6.6
Fungi	12×10^5	1×10^5 **	12.0
Protozoa	24×10^2	10×10^2 **	2.4
Algae	5×10^3	27×10^3 *	0.2
NUTRITIONAL GROUPS OF BACTERIA			
Ammonifiers	500×10^6	4×10^6 **	125.0
Gas-producing anaerobes	39×10^4	3×10^4 *	13.0
Anaerobes	12×10^6	6×10^6 *	2.0
Denitrifiers	126×10^6	1×10^5 **	1260.0
Aerobic cellulose decomposers	7×10^5	1×10^5 *	7.0
Anaerobic cellulose decomposers	9×10^3	3×10^3 NS	3.0
Spore formers	9×10^5	6×10^5 NS	1.5
Azotobacter	17×10^6	$< 1 \times 10^4$ **	—

†** = Significantly different at the 1% probability level
* = Significantly different at the 5% probability level
NS = Not significant

of the growing root. Rhizoplane fungi are dominated by relatively few genera such as *Fusarium* and *Rhizoctonia*. These changes in numbers are illustrated by the data in Table 6.5. Protozoa also increase in the rhizosphere, presumably in response to the greater number of prey, but they only become important after the rise in bacterial numbers has taken place (Fig. 6.6). Chromogenic bacteria often show much greater increase in the rhizosphere than colourless forms, though the reason(s) for this is unknown. Figure 6.6 also shows the effect of plant age on the rhizosphere population. Often it is not the chronological but the physiological age that is important, numbers reaching a maximum during the early stages of flowering.

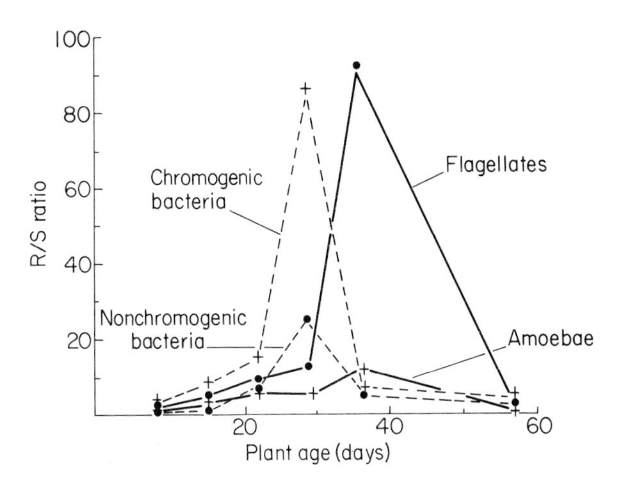

Figure 6.6. The rise in bacterial and protozoan numbers in the rhizosphere of *Sinapsis alba* L. (white mustard). The R:S ratio (Table 6.5) is plotted against plant age, the maximum rhizosphere effect for protozoa is later than that for bacteria. (Modified from data of J.F. Darbyshire and M.R. Greaves, 1967. Protozoa and bacteria in the rhizosphere of *Sinapsis alba* L., *Trifolium repens* L., and *Lolium perenne* L. *Canadian Journal of Microbiology* **13**, 1057–1068.)

If, instead of considering taxonomic groupings, nutritional groups are studied the greatest increase is shown by organisms with fairly complex requirements, particularly for amino acids (Tables 6.5 and 6.6). This reflects the common occurrence of amino acids in the exudate.

The extent of the rhizosphere depends not only on the exudate but also on the physical and chemical factors in the soil, which affect the diffusion of substances, and on the numbers of micro-organisms present. In a soil very low in organic matter any addition will have a great effect and conversely if the available soil organic matter is high then a little more from the root will have less effect. If there are very high microbial populations on the rhizoplane then these will use more or less all the exudate. A plant producing a lot of exudate in pure

Table 6.6 Nutritional groups in the rhizosphere compared with their abundance in control soil. (From A.G. Lochhead and R.H. Thexton (1947) Quantitative studies of soil micro-organisms. VII. The 'rhizosphere effect' in relation to the amino acid nutrition of bacteria. *Canadian Journal of Research* **25C**, 20–26.)

Nutritional requirement for maximum growth	Millions per g		
	Rhizosphere	Control	R:S ratio
Basal medium	119.7	4.5	16.6
One or more amino acids	133.0	2.6	51.2
Growth factors	79.8	8.7	9.2
Amino acids and growth factors	79.8	6.1	13.1
Yeast extract	62.2	6.1	10.2
Soil extract	30.8	2.6	11.8
Yeast and soil extracts	13.3	4.2	3.2
'Total' plate count	532.0	37.5	14.2

sand with very few micro-organisms may have a rhizosphere extending several centimetres: a plant producing moderate or small amounts of exudate in a soil rich in organic matter with a high rhizoplane population may have a rhizosphere almost coincident with the rhizoplane, the distance measured from the root being a fraction of a mm or just a few micrometres. The greatest stimulation of numbers is on the rhizoplane and the abundance decreases to normal soil levels as the distance from the root increases. Different species of microbes vary in their distribution, a few being confined to the rhizoplane, others having a more general distribution in the rhizosphere and yet others as we have already seen showing no rhizosphere effect or decreasing in numbers (Table 6.7).

Table 6.7 The effect of distance from the root surface on rhizosphere micro-organisms. (From G.C. Papavizas and C.B. Davey (1961) Extent and nature of the rhizosphere of *Lupinus*. *Plant and Soil* **14**, 215–236.)

	Millions/g dry wt.		Thousands/g dry weight					
Distance from the root mm	Bacteria	Strepto-mycetes	Fungi	Cylindro-carpon radicicola	Paecilo-myces marquandi	Fusarium oxysporum	Tricho-derma viride
Rhizoplane = 0	159.0	46.7	355	4.9	9.0	5.3	5.9
0 to 3	49.7	15.5	176	0	2.8	2.5	22.1
3 to 6	38.0	11.4	170	0	1.6	2.8	6.9
9 to 12	37.4	11.8	130	0	1.5	4.5	14.9
15 to 18	34.2	10.1	117	0	0	4.1	7.3
Control = 80	27.3	9.1	91	0	0	5.1	11.1

Influences on the microflora are not always as simple to interpret as these examples. Decreasing light intensity would be expected to decrease photosynthesis and therefore the amount of exudate and so there would be fewer organisms in the rhizosphere. This relationship can be shown for some fungi (e.g. *Trichoderma* and *Gliomastix*, Table 6.8) but there are others that increase, particularly some of the parasites such as *Rhizoctonia*, as the light intensity decreases. There are two possible explanations for this: at these very low intensities the resistance mechanisms of the plant may be poor or inoperative, or there may be antagonism on the root surface (e.g. with *Trichoderma viride*) and as the antagonist decreases, in response to the light intensity, the parasites increase. Some of the fungi listed in Table 6.8 are not affected by the light intensity within the range used.

Microbial activity in the rhizosphere, as opposed to numbers, is difficult to measure. However the microbial respiration rate per unit volume of soil is estimated to be greater than in control soils. Free-living nitrogen fixation in soil occurs mostly in the rhizosphere, especially on the rhizoplane. *Azotobacter paspali* specifically colonizes the mucigel (see below) of *Paspalum* and is very much stimulated. A bacterium tentatively identified as *Spirillum lipoferum* is associated with the roots of tropical grasses (e.g. *Digitaria decumbens* and some strains of *Zea mays*) and fixes significant amounts of nitrogen. Ammonification (breakdown of amino acids particularly) and nitrification are increased in the rhizosphere

Table 6.8 The effect of decreased illumination on the frequency, per hundred root segments, of surface fungi on the roots of beech seedlings. (Selected from J.L. Harley and J.S. Waid (1955) The effect of light upon the roots of beech and its surface population. *Plant and Soil* 7, 96–112.)

	% daylight radiation				
Genus	25.1	14.2	10.6	6.1	3.6
Trichoderma	72	43	45	22	36
Gliomastix	21	17	12	7	12
Verticillium	4	8	5	5	8
Gliocladium	7	6	10	5	6
Cylindrocarpon	25	31	51	59	44
Rhizoctonia	1	19	22	32	23
Penicillium	39	34	49	43	58

but, unfortunately for the plants, the rate of denitrification also rises as would be expected from the increase in anaerobic microhabitats (see Chapter 4, page 46) caused by the general increase in microbial activity. There is also microbial immobilization of nitrogen in the rhizosphere.

It is probable that the distribution patterns of micro-organisms on soil crumbs in the rhizosphere are similar to that in the bulk soil (see above, page 66). On the rhizoplane however the situation is somewhat different, for the root, and the bacteria growing on it, produce mucilage (the mucigel) which can be seen quite clearly on direct examination (Fig. 6.7A). Micro-organisms are embedded in the mucilage either as single cells or as microcolonies. The root surface, despite the apparently high numbers recorded, is rather sparsely populated by micro-organisms in terms of the amount of surface actually occupied: 10% cover would be considered dense and it may commonly be only 1 to 4%. There are places, such as where microcolonies occur, where the population is much greater and there may even be several layers of bacteria (Fig. 6.7B and C). Some workers report a more or less random distribution but other results suggest that the distribution is related to the microtopography of the root, microbes being concentrated along the epidermal cell boundaries, possibly where the exudation is greatest.

The effect of the micro-organisms on the plant

The shoot growth, seed germination and the time of flowering may be affected by changes in plant hormone concentrations, though whether hormones are produced by the microbes or whether the production and distribution of the plants own hormones is changed remains in doubt. The case of gibberellin production by *Azotobacter* has already been mentioned (Chapter 4, page 47) and indoleacetic acid is produced by *Rhizobium* during penetration (page 46). It is a common experience that plants grow poorly in sterilized soil; though this effect is usually attributed to toxins produced during sterilization, changes in hormone balance could also be a factor. In one study up to 50% of rhizosphere and rhizoplane organisms produced plant growth promoters in culture, but a similar number of isolates (not necessarily the same ones) produced growth inhibitors

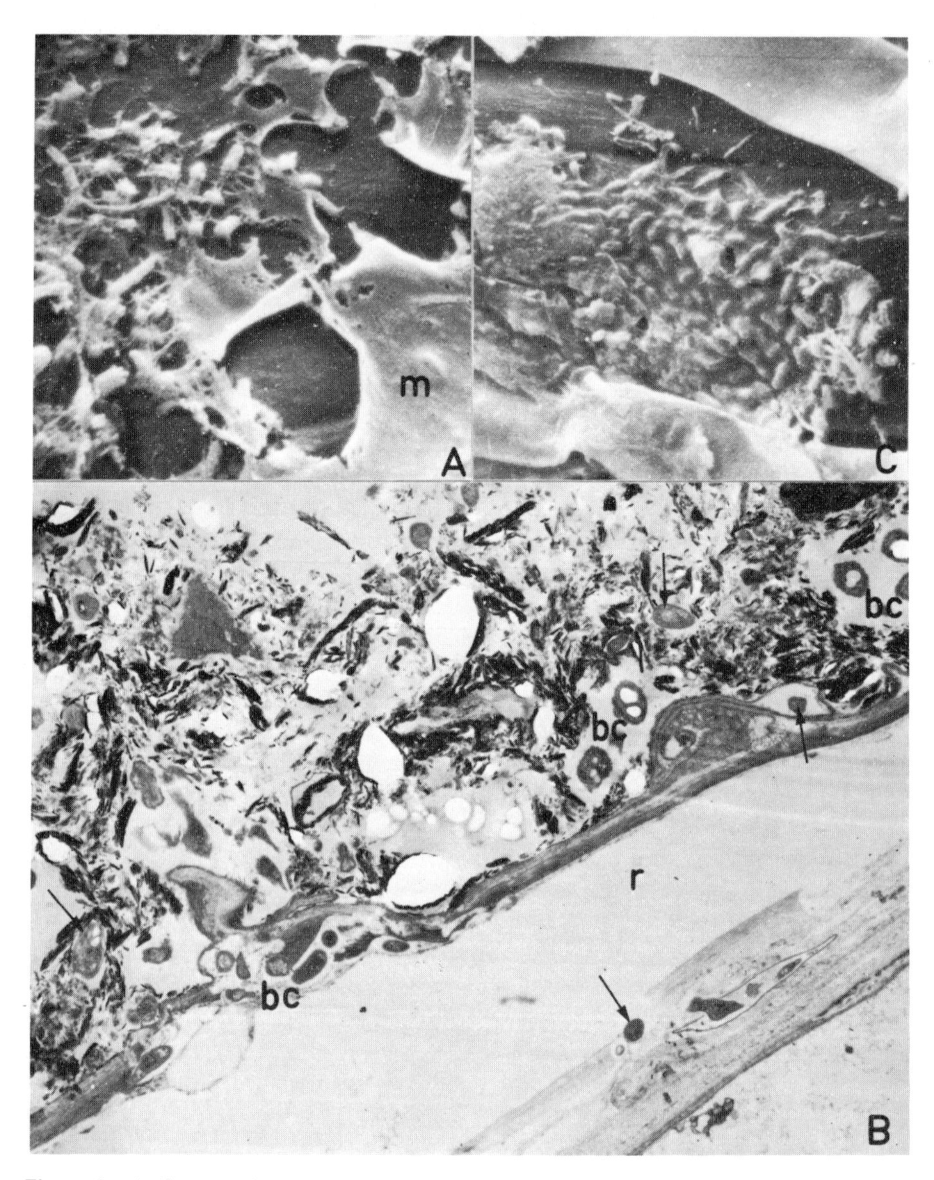

Figure 6.7. A. 'Stereoscan' photograph of a wheat root with smooth sheets of mucilage (m; of plant origin?) and the more fibrous material associated with the bacteria (to the left of the picture). ×2,100. (From A.D. Rovira and R. Campbell (1975). Scanning electron microscopy of micro-organisms on the roots of wheat. *Microbial Ecology* **1**, 15–23). B. Transmission electron micrograph of *Loasa vulcanica*: the root (r) is in the lower right of the picture and the cortical cells have bacteria (arrow) within them. Individual bacteria (arrows) and bacterial microcolonies (bc) occur in the soil. Clay particles appear as electron opaque, lamellate structures. ×6,100 (Unpublished micrograph by Dr A. Beckett, Bristol University). C. 'Stereoscan' photograph of a bacterial microcolony on a wheat root ×1,400. (From A.D. Rovira and R. Campbell (1975) Ibid.).

under these conditions. Sterile soil inoculated with microbes results in stunting of root systems, a reduction in the number of lateral roots and a delay in root hair formation when compared with axenic plants. These changes in root morphology could have a profound effect on nutrient uptake, purely from consideration of the volume of soil exploited. This would be particularly important for such immobile compounds as phosphate where zones of depletion are rapidly established around roots.

The best known change in root morphology is that caused by the symbiotic associations with the plant. Nitrogen fixing nodules have been discussed already (page 42) and mycorrhizae mentioned in connection with phosphorus uptake (page 59). There are two main types of mycorrhizae, endotrophic (including vesicular arbuscular and orchid types) and ectotrophic (further reading 7). The orchid mycorrhizae are of interest from a plant nutritional and academic point of view but are not of great economic importance. They may be necessary, or at least beneficial, in seed germination and in the non-photosynthetic protocorm the hyphae occur within the cells (Fig. 6.8A) and provide the orchid with available carbohydrate and probably vitamins and hormones. In culture the fungi are capable of utilizing insoluble carbohydrates such as cellulose whereas the protocorm is dependent on an exogenous supply of simple sugars. Mature orchids are mostly able to produce sugars etc. by ordinary photosynthesis. Sometimes, particularly in the seed germination and seedling stage, the fungus may parasitize the orchid: the relationship seems to be rather unstable. The fungi in the association are often *Rhizoctonia* spp. which are normally root parasites of plants, and the orchids have somehow turned the usual relationship about: the way in which this is done is not known, though the phytoalexin orchinol has been implicated. It is an interesting example of variation in the balance of virulence and resistance between parasite and host, and of how different sorts of symbiosis may grade into one another.

Vesicular arbuscular mycorrhizae are associations between species of phycomycetes, usually called *Endogone*, and a great number of Angiosperms. The aseptate hyphae penetrate the cortical cells and form intricately branched organs called arbuscules; larger hyphae also occur which may have terminal or intercalary swellings and vesicles (Fig. 6.8B). Vesicular arbuscular mycorrhizae are very common on many species of plants and are widely reported from all parts of the world. They have been shown to increase nutrient uptake and translocation, particularly of phosphorus in deficient soils. They are most probably very important in plant nutrition, in both natural communities and agricultural situations, but until recently they have not been so well studied as ectotrophic mycorrhizae.

Ectotrophic mycorrhizae are common on many forest trees, particularly pines, beech and birch in the north temperate region. They form a sheath over the outside of the roots and penetrate between the cortical cells, to form the Hartig net, though they do not usually occur within cells (Fig. 6.8D). The root morphology is changed by the thickness of the sheath and by the repeated formation of short branches with blunt tips and limited growth (Fig. 6.8C). The precise pattern of the branching and the amount of mycelium ramifying into the surrounding soil varies with the species of tree and the fungus. Any one tree species

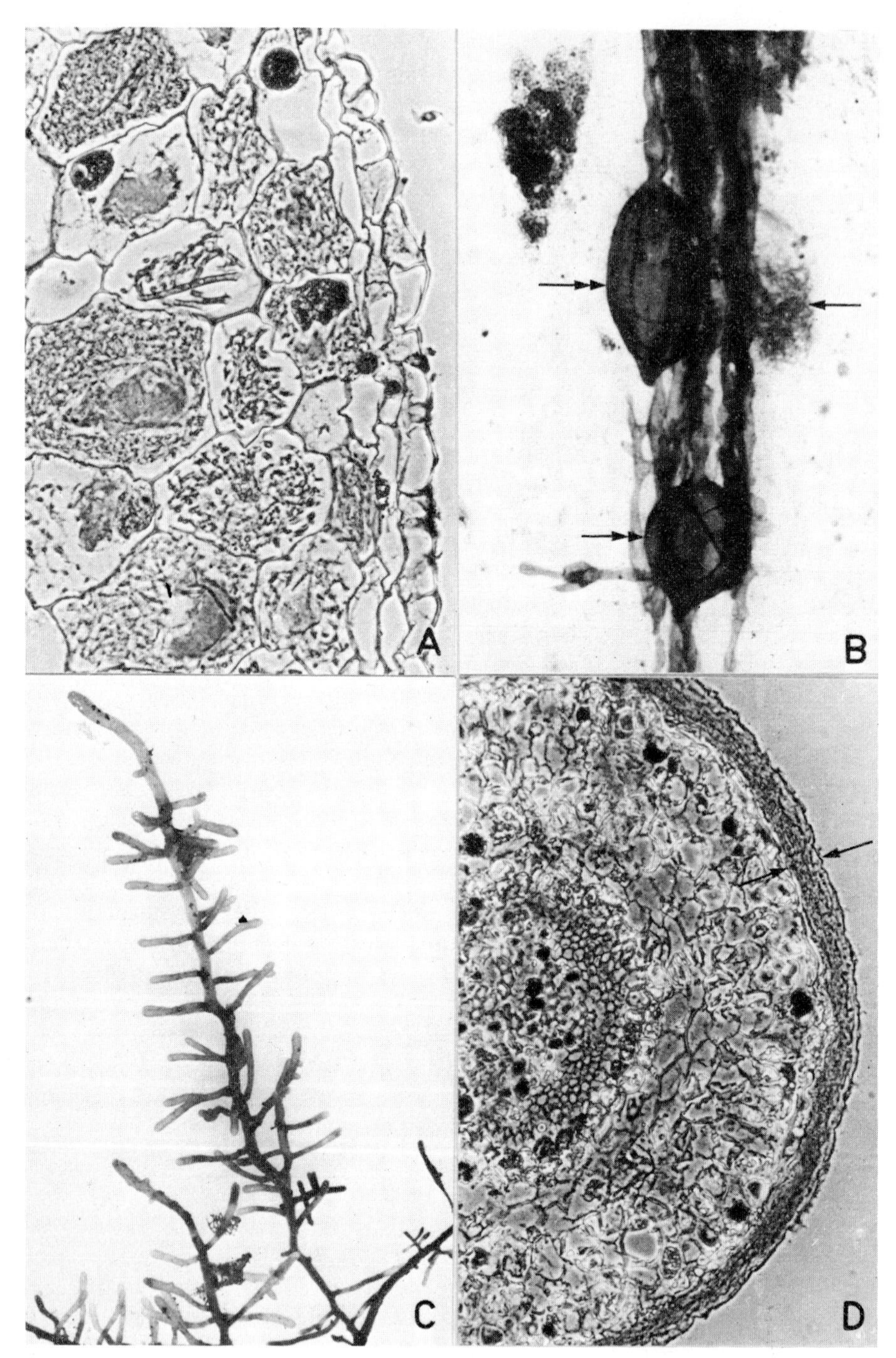

Figure 6.8. Mycorrhizal associations. A. Endotrophic: hyphae within the cells of a protocorm of the orchid *Neottia*. × 200. B. Endotrophic: vesicular arbuscular: on *Plantago* with intracellular hyphae, vesicles (double arrows) and an arbuscule (arrow). × 400. C. Ectotrophic: beech showing the growth form of the infected roots. (approx. natural size). D. Ectotrophic: hyphae penetrate between the cortical cells and form a sheath (between arrows) around the root of a beech tree. × 200.

may form mycorrhizal relationships with many different sorts of fungi and conversely most fungi (e.g. *Cenococcum*) have a wide host range, though some (e.g. *Boletus elegans*) are apparently restricted to a single host genus (*Larix*). Common mycorrhizal genera are basidiomycetes, particularly Agaricaceae such as *Boletus, Amanita, Tricholoma, Russula* and *Lactarius*. Many of these fungi, or at least their mycorrhizal strains, do not have a great ability to break down complex carbohydrates in culture (cf orchids). They presumably receive most, if not all, of their carbohydrate from the host tree and they convert it into mannitol, trehalose and glycogen which the tree cannot utilize, so the movement of carbohydrate is in one direction only. The tree has improved growth, mostly because of increased mineral nutrient uptake; phosphate in particular accumulates in the sheath and may later be transferred to the tree. Improved uptake of potassium and nitrogen has also been shown. Increased uptake may be due to several factors. The branching and thickening of the root increases the surface area and the hyphal strands further extend the soil volume penetrated so that more of the soil's pool of immobile elements like phosphorus is utilized; the mycorrhizal fungus may be more efficient than the root at uptake, and lastly the mycorrhizal roots live longer and are less susceptible to disease than uninfected roots. The relationship is therefore mutualism. The maximum development of mycorrhizae occurs in soil of moderate to low fertility and when trees receive full light. Ectotrophic mycorrhizae have a rhizosphere of their own, and the R : S ratio is usually greater than uninfected roots. The young root is colonized by the mycorrhizal fungus in competition with the normal rhizoplane microflora and once the sheath is formed the root exudates are modified by the fungus.

Similar effects on nutrient uptake have been reported for the rhizosphere population in general, though the evidence is much more in dispute than that for mycorrhizae. It is plausible that nutrient uptake could be affected because everything that reaches the root has had to pass through the rhizosphere. Apart from phosphorus and nitrogen (see pages 59 and 67) the uptake of the following elements has been reported to be affected by micro-organisms in the rhizosphere of laboratory plants: increased uptake for potassium, zinc, calcium, magnesium, iron, manganese, molybdenum and aluminium, and decreases for rubidium and strontium. The results for sulphur are particularly inconsistent and reverse responses to those indicated are sometimes reported for all those elements listed above. It is probable that response varies with the plant and the micro-organisms in a particular environment and there will have to be much further work before these effects are understood. The most likely mechanisms are the alteration of pH, the dissolving of previously insoluble materials with metabolic acids and the release of minerals during decay of organic matter.

BIOLOGICAL CONTROL OF SOIL-BORNE PATHOGENS

Plant pathogens in the soil must be able to survive in the absence of their hosts. They may compete with saprophytes or they may remain dormant until a suitable host is near, as do some *Fusarium* species whose dormancy is broken by root exudates from a susceptible host. Root exudates may also attract the motile zoospores of *Phytophthora* which congregate on the susceptible roots in the exudation zone. Once the parasite is inside its host the microbial competition is reduced: the ability of successful parasites to penetrate a normally defended position gives them the benefit of living in a 'sheltered environment'.

Plant pathogens, which are mostly fungi in the soil, react in the same way as other micro-organisms to the environmental factors already discussed. They are antagonized and sometimes antagonistic; hyphae, spores and sclerotia may be lysed by other soil organisms. The essence of biological control of soil-borne pathogens (further reading 2) is to tip the balance away from the pathogen and in favour of the saprophytic soil flora. This flora is most probably important in pathogen control for disease is often much more severe or the infection rate is increased if other microbes are removed. Soil micro-organisms may play a part in the specificity of host and pathogen.

Methods of biological control either increase the general microbial activity

Table 6.9 The genetic control of wheat rhizosphere and the apparent effect on the pathogenicity of *Cochliobolus sativus*. Variety Apex is resistant, S-615 is susceptible and S-A5B is S-615 with the substitution of the Apex 5B chromosome. Data within a column followed by the same letter do not differ significantly (P=0.05). (From J.L. Neal, T.G. Atkinson and R.I. Larson (1970) Changes in the rhizosphere microflora of spring wheat induced by disomic substitution of a chromosome. *Canadian Journal of Microbiology* **16**, 153–158. See also in G.W. Bruehl (Ed.) (1975) *Biology and Control of Soil-borne Plant Pathogens*. American Phytopathological Society, St Paul, Minnesota. pp. 216.)

		Rhizosphere	
Variety	Resistance	Bacteria Millions/g dry weight	Fungi Thousands/g dry weight
Apex	Resistant	251.5 a	342.9 a
S-615	Susceptible	576.8 b	123.7 b
S-A5B	Resistant	266.5 a	81.5 c
No plant	—	44.9 c	119.5 b

in the soil so that there is more antagonism and lysis during saprophytic survival of a pathogen, or they try to prevent infection by manipulating the rhizosphere flora itself. The latter is under the genetic control of the host and some of the plant breeding programmes for disease resistance have changed the rhizosphere at the same time as resistance has been conferred (Table 6.9).

Soil microbial populations are relatively stable and it is difficult to change their composition unless the environment is altered in some way. The ploughing in of green manure crops, especially legumes, greatly increases microbial activity and has been used in control of many diseases e.g. *Phymatotrichum* root-rot of cotton, potato scab (*Streptomyces scabies*). Similarly the control of pH by additions of sulphur or the correct balance of nitrogenous fertilizers may sometimes favour antagonists more than the pathogen, as well as directly affecting the pathogen's growth. The most drastic measure to alter the soil population is sterilization, or more usually partial sterilization, which is routinely performed in glasshouses and other places with relatively limited quantities of soil. Various substances have been used including steam, aerated steam, formalin, chloropicrin, methylbromide and many proprietary formulations. The aim is to kill the pathogen but leave as much as possible of the rest of the soil flora alive. If the soil is left sterile, or with a seriously depleted saprophytic flora, and the pathogen is then reintroduced by accident, there will be very serious disease, but at correct treatment levels there is a very rapid rise in the numbers of the remaining soil saprophytes e.g. *Pseudomonas* and *Bacillus*, to fill the vacant niches and to utilize all the dead remains. This is a good, if extreme, example of the zymogenous response of micro-organisms. One of the commonest and quickest fungal recolonizers is *Trichoderma*, a well known general antagonist which may help to prevent renewed growth of pathogens. Successions have been described for partially sterilized soil and modifications of the flora may still be detectable many months after the treatment. There is now considerable practical expertise on the best levels of the particular fumigants to use for particular problems.

Having disturbed the equilibrium by some means the opportunity may be taken to introduce into the soil some alien antagonist which has been specially isolated. There are now well defined ways of obtaining such antagonists (further reading 2) and they are usually applied as a seed dressing or as a dust or drench. Combining an imposed change in the environment with inoculation can have a dramatic effect on disease severity. There have also been reports of yield increases in excess of those expected from disease control, and microbial effects on auxin and/or gibberellin have been suggested as a possible cause.

Some soils are initially favourable for the pathogen and disease increases, particularly with continuous monoculture, to serious levels. The usual practice has been to break the monoculture for a few years to allow the inoculum potential of the pathogen to decrease. Some diseases eventually decline if the same crop is repeatedly grown, because the equilibrium between the parasite and antagonistic micro-organisms in the soil is restored. This situation occurs in take-all disease (*Gaeumannomyces graminis*) of wheat where the build-up takes about three years and the disease declines to low levels about five years after the start of monoculture. Many antagonistic organisms, and viruses, that attack the fungus have been found in the soil during the decline period and through the subsequent low levels of the disease.

Biological control of soil pathogens has usually been successful when rather gross changes are made in the environment to encourage resident antagonists, but as the interactions in the complex soil environment are better understood there will be some chance of designing control methods rather than the somewhat empirical procedures now used. Most probably a system of integrated control based on chemical or cultural measures followed by the biological exploitation of the disturbed equilibrium will be most successful.

THE EFFECT OF MAN ON SOIL MICRO-ORGANISMS

Apart from the attempts to change the soil flora in biological control, the effect of man on the nitrogen and carbon cycles has already been discussed (pages 46 and 32). In general, agricultural practices such as ploughing, liming, fertilizing etc. increase the microbial activity in the soil. There is some apprehension that modern methods such as direct sowing after using paraquat, rather than ploughing, and the burning of straw and stubble, rather than its return to the soil enriched in nitrogen as farmyard manure, may in the long run be bad for soil structure and fertility because of the lack of organic matter left to become humus. Rothamsted experimental station in England has had trials running on the problem for about 120 years and there are decreases in organic matter in soils given continuous fertilizer treatments rather than organic manures but the differences are not great. These time spans may be too short for a correct assessment, for the mean residence time for humus fractions in some grassland soils has been determined by carbon dating at up to 1500 years. When organic matter is returned to the soil from modern intensive animal units it is often in the form of a slurry rather than a solid manure. The high organic content and the excess liquid may cause anaerobic conditions which can be aggravated by the blocking of the soil pores with the finely divided organic matter. Urban sewage or partially treated sludge is also spread on land sometimes and here there are additional problems because of the heavy metal content. This is not serious unless repeated applications are made. Heavy metal contamination is also a problem in composting urban waste (page 29).

The variability of micro-organisms and the complexity of the interactions between the soil components, which result in the stability of the soil equilibria have, in most cases, allowed drastic changes in vegetation types and environmental conditions to be made by agriculture and industry without a breakdown of the soil system occurring.

FURTHER READING

1 ALEXANDER M. (1971) *Microbial Ecology*. Wiley, New York and London. pp. 511.
General text.

2 BAKER K.F. & COOK R.J. (1974) *Biological Control of Plant Pathogens*. Freeman, San Francisco. pp. 433.
Competition and antagonism between plant pathogens and soil saprophytes.

3 BOWEN G.D. & ROVIRA A.D. (1976) Microbial colonization of plant roots. *Annual Review of Phytopathology* **14**, 121–144.
Review of the rhizosphere.

4 BURGES A. & RAW F. Eds. (1967) *Soil Biology*. Academic Press, London and New York. pp. 532.
Includes algae, protozoa and soil animals.

5 DICKINSON C.H. & PUGH G.J.F. Eds. (1974) *Biology of Plant Litter Decomposition*. Vols 1 and 2. Academic Press, London and New York. pp. 146 and 175.
Includes the microbiology of decomposition.

6 GRAY T.R.G. & WILLIAMS S.T. (1971) *Soil Micro-organisms*. Oliver and Boyd, Edinburgh. pp. 240.
General text.

7 HARLEY J.L. (1969) *The Biology of Mycorrhiza*. Leonard Hill, London. pp. 334.
The basic text on mycorrhizae.

8 NOLAND L.E. & GOJDICS M. (1967) The ecology of free-living Protozoa. In T-T. Chen. Ed. *Research in Protozoology* Vol. 2. Pergamon Press, Oxford. pp. 215–266.
Includes a review of soil protozoa.

9 PARKINSON D., GRAY T.R.G. & WILLIAMS S.T. (1971) *Methods for Studying the Ecology of Soil Micro-organisms*. IBP Handbook 19. Blackwell, Oxford and Edinburgh. pp. 116.
Includes all the main methods.

10 ROUND F.E. (1977) *Algal Ecology*. (In press.)
Includes soil algae.

11 Sandon H. (1927) *The Composition and Distribution of the Protozoan Fauna of the Soil.* Oliver and Boyd, Edinburgh and London. pp. 237.
Comprehensive cover of Protozoa.
12 Stotsky G. (1974) Activity, ecology and population dynamics of micro-organisms in soil. In A.I. Laskin and H. Lechevalier. Eds. *Microbial Ecology.* CRC Press, Cleveland, Ohio. pp. 191.
Surface adsorption effects in particular.
13 Swift M.J. (1976) Species diversity and the structure of microbial communities in terrestrial habitats. In J.M. Anderson and A. Macfadyen. Eds. *The Role of Terrestrial and Aquatic Organisms in Decomposition Processes.* Blackwell, Oxford, London, Edinburgh and Melbourne. pp. 185–222.
A good review of community structure and successions.
14 Walker N. Ed. (1975) *Soil Microbiology.* Butterworths, London and Boston. pp. 262.
Reviews selected aspects of the distribution and activities of soil micro-organisms.

7 The structure and dynamics of microbial populations in water*

Water never occurs naturally in the pure state; as we have seen rain contains various compounds washed out of the atmosphere and during percolation through the soil and rocks further materials are dissolved (Chapters 3, 4 and 5). Rather more than 70% of the earth's surface is covered with water and 84% of this has a depth greater than 2000 m. Only 1% of the area is freshwater, but because of its accessibility more is known about this small fraction than about the oceans. Many different categories of fresh water have been described (further reading 3), based on their nutrient status and other chemical and physical characteristics.

We will consider first the environmental factors (further reading 3, 5 and 9) which influence the growth and distribution of micro-organisms in aquatic habitats and then the different communities that occur (further reading 3, 4 and 10).

WATER AS AN ENVIRONMENT

Temperature and water density

Water is physically and chemically a rather unusual material, for the molecules associate in hydrogen bonded chains in a semi-crystalline structure, particularly at temperatures near to 0°C. This results in peculiar density changes, the maximum being at 4°C. There is a variation in temperature with depth in lakes and oceans and also seasonal changes outside the tropics. In spring cold lake water in the temperate zone is heated by the sun, creating a less dense surface layer, the epilimnion, which floats on a colder, more dense layer, the hypolimnion. The transition zone between the two layers, the thermocline, can be quite sharply defined. This stratification persists in deep lakes throughout the summer, though shallow waters may be unstratified. In autumn radiation of heat from the surface cools the epilimnion until it is the same temperature as the underlying hypolimnion, so the two layers mix. This autumn overturn, which is assisted by wind turbulence, ends the summer stratification (Fig. 7.1). Stable stratification can form in winter, but the temperature relations are reversed so that the water at or near 4°C lies at the bottom with colder, less dense, water above. An ice cover stabilizes the situation further and prevents turbulent mixing (Fig. 7.1). This stratification breaks down as the surface water warms to 4°C again in spring, resulting in the spring overturn, but the winter stratification may not occur if the temperatures are not low enough or if the water is mixed by winds.

* I would particularly like to thank Dr F.E. Round, Department of Botany, Bristol University for the help and advice that he has given during the preparation of this chapter.

Summer stratification occurs in temperate oceans too, though the terms hypolimnion and epilimnion are replaced by deep and surface waters respectively. Polar oceans do not have thermoclines, but in the tropics they may be permanently present. Tropical lakes are more complicated; many do not appear to be thermally stratified, others develop stratification on a diurnal basis, and yet others may have more stable stratification which may occasionally break down, but not in a regular seasonal cycle.

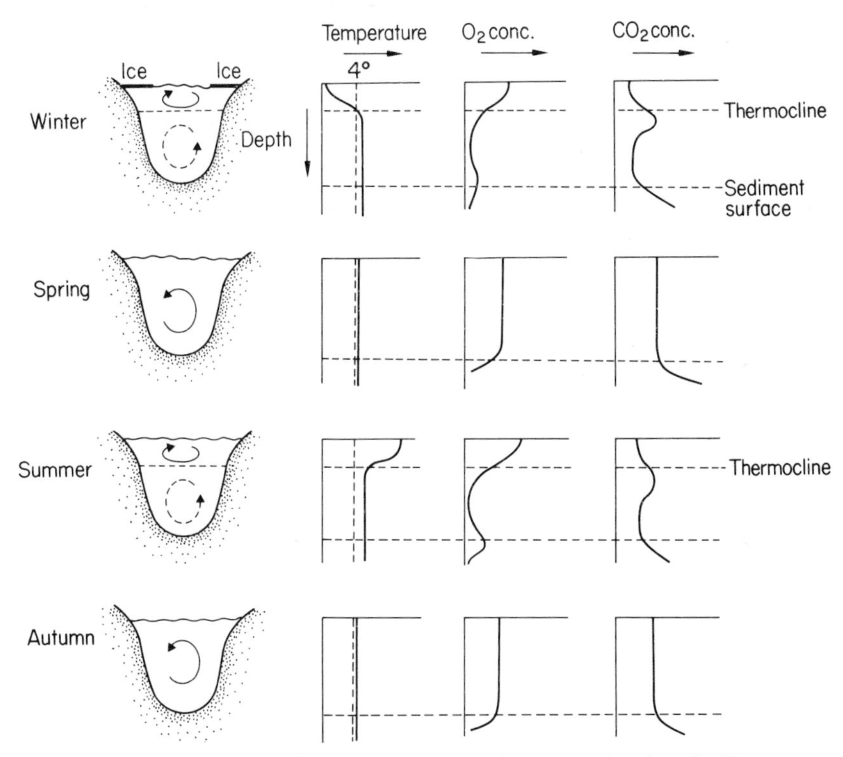

Figure 7.1. Diagram of the variation of temperature, and oxygen and carbon dioxide concentrations with depth at different times of year under different systems of stratification. Winter stratification does not occur in some lakes or in most seas, in which case there is a continuous period of mixing from the autumn overturn to the onset of summer stratification.

Thermoclines, particularly if they are permanent are very important in controlling the productivity of the aquatic ecosystem. With a stable thermocline there is very limited mixing between the upper and lower water layers, though each may have a circulation within itself. The photic (light receiving) zone in the epilimnion, in which photosynthesis occurs, can become depleted in nutrients while just below in the hypolimnion they occur in relative abundance. This is one of the reasons for the relatively low productivity of most tropical oceans. Where mixing does occur in the tropics, for example on those coasts where there is an upwelling from nutrient rich deep water, the productivity can be very high. The high productivity of polar regions during their summer is partly due to the lack of a thermocline so that all the nutrients in the water column are potentially available.

Rivers are usually too well mixed for there to be any variation of temperature with depth, but the temperature fluctuations in the whole water mass are often large because of the large surface area to volume ratio. Conversely the temperature fluctuations found beneath the surfaces of large lakes and the oceans are very small: the range over all the world's oceans is only $-2°C$ to $40°C$ with the majority of sea water (90% by volume) permanently below 5°C. Seasonal temperature changes only affect the top 10 to 200 m of oceanic water and even here the range is usually less than 10°. Diurnal changes are even smaller, 0.2 to 0.3°C at the surface layers. The greater part of the biomass in water is however in the comparatively shallow photic zone where temperatures are more moderate.

Dissolved gases and pH

Water is slightly dissociated (the K_W being 10^{-14} at pH 7.00) to form the hydroxonium (H_3O^+) and hydroxyl (OH^-) ions. The hydrogen ion itself does not exist in water though it is commonly referred to, and will be here, instead of hydroxonium. Compounds which dissolve to produce hydrogen ions greatly affect the pH and by far the most important of these in natural waters is carbon dioxide which is very soluble (further reading 3 and 10). There are two possible reactions:

$$CO_2 + H_2O \rightleftharpoons H_2CO_3$$
$$CO_2 + OH^- \rightleftharpoons HCO_3^-$$

The H_2CO_3 is itself strongly dissociated:

$$H_2CO_3 \rightleftharpoons H^+ + HCO_3^- \rightleftharpoons H^+ + CO_3''$$

These reactions are all in equilibrium so the concentration (or activity to be correct) of the species of inorganic carbon will determine the pH, and conversely if something outside the equilibrium alters the pH then the proportions of the different carbon species will change (Fig. 7.2). The removal of carbon dioxide or bicarbonate during photosynthesis raises the pH and conversely the addition of carbon dioxide from respiration decreases it. Acidic water containing carbon dioxide will dissolve calcium carbonate (limestone etc.) as calcium bicarbonate but this may be reprecipitated as carbonate again if carbon dioxide is removed and the pH rises:

$$Ca(HCO_3)_2 \rightleftharpoons CaCO_3 + H_2O + CO_2$$

The equilibria can therefore become quite complicated and furthermore there are connections with other cations in the water, particularly magnesium whose carbonate is soluble.

The vertical distribution of carbon dioxide in lakes and the sea is closely linked, inversely, with oxygen levels (Fig. 7.1). Carbon dioxide concentration is low at the surface and increases with depth, often not uniformly because of stratification. Photosynthesis causes fluctuations in the concentration, lowering it during light periods.

The amount of oxygen diffusing into water depends on the partial pressure in the air above, the temperature and the salinity. A saturated solution at 16°C and normal atmospheric pressure contains 5.9 ml/l in sea water and 7.2 ml/l in fresh water. Near the surface the water may be supersaturated with oxygen because of photosynthesis but the concentration decreases with depth, reaching a minimum below the photic zone due to the respiration of bacteria whose numbers frequently reach a maximum there. At depths below 600–900 m in the oceans the oxygen concentration may rise again in the nutrient-poor waters and then finally decrease at and just below the sediment surface.

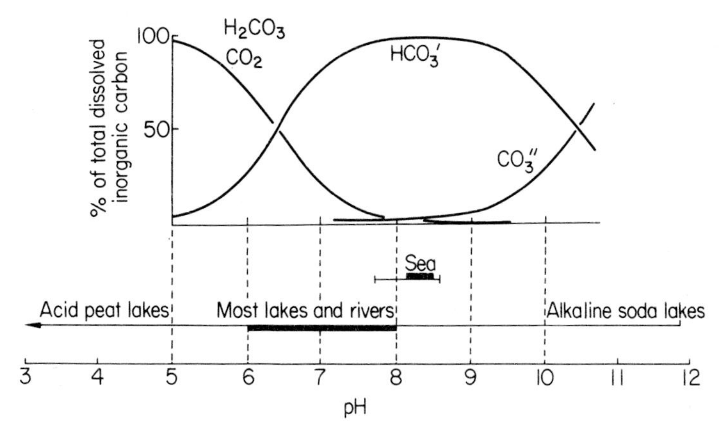

Figure 7.2. The relationship between the pH and the species of inorganic carbon present in water. The usual pH ranges and the possible maxima and minima found in natural waters are shown. The curves are for fresh water, those for sea water are slightly different. (Modified from J.P. Riley and G. Skirrow (Eds.) (1965). *Chemical Oceanography*, Volume 1. Academic Press, London and New York, pp. 712.)

Rivers and other flowing waters usually have a more even distribution of oxygen and carbon dioxide. In fast, clear streams their levels may be close to saturation but if there is little turbulent mixing or the organic matter content is high (either because of sewage or the natural addition of leaves etc.) then the oxygen levels may be low; these latter conditions are characteristic of slow flowing, lowland rivers and streams.

In dark, anaerobic environments e.g. many organic sediments, only some bacteria and a few protozoa such as *Pelomyxa* sp. and *Metopus es* can actively grow and specialized communities may be associated with the production of hydrogen sulphide by *Desulfovibrio desulfuricans* (see below page 118).

Salinity

Sea water usually contains about 3.5% salt (w/v), though it varies slightly in different parts of the oceans and there are relatively enclosed areas like the Red Sea where the salinity may be higher and inland lakes (e.g. the Dead Sea) where very high values are found. A clear distinction can usually be made between the flora and fauna of fresh and salt water. There are several possible responses that a micro-organism may make to saline water: it may be intolerant and die, and this is the reaction of most fresh water species particularly if the change to full salinity is abrupt. Secondly the microbe may tolerate a range of salinities (i.e. are euryhaline) but grow better at low levels. Many marine fungi, some protozoa and many flagellates and dinoflagellates are in this category, growing slowly in pure sea water. The third alternative is that there is a requirement for saline conditions, either for the sodium ion or for osmotic reasons. Such organisms are obligate halophiles and most marine bacteria, algae and protozoa are in this category, and some (e.g. *Peridinium balticum*) will only grow within a narrow salinity range and these are termed stenohaline. Salinity affects the responses of micro-organisms to pressure and temperature and also alters the solubility of gases.

Sunlight

The amount of sunlight reaching the surface affects the temperature and the productivity of aquatic environments: there are of course seasonal and diurnal variations in the amount of radiation received. Particulate matter will scatter the light and in some estuaries with extremely turbid water penetration may be only a few centimetres, but in general 50% of the light reaches 3 or 4 m in inshore waters. In the purest lake waters the photic zone may extend to 200 m but usually only about 10% of the light reaches 45 m even in apparently clear water (further reading 3 and 5). Apart from the decrease in the quantity of light with depth there is also a change in the colour, blue light (short wavelengths) being transmitted most and red light the least.

Light, though it is obviously essential to the photoautotrophs, may also be harmful to micro-organisms. It has been claimed to be bacteriocidal and this has been given as one of the reasons for the death of bacteria, especially those from sewage, in sea water. The effect is however probably minimal and only operates in shallow water (< 20 m). High light intensities may affect bacterial activity without killing, e.g. reductions occur in the rates of the oxidation by *Nitrosomonas* and *Nitrobacter* and also in denitrification.

Cell division may be related to day length or the dark period. Thus the diatom *Nitzschia* divides mostly in the light, while the dinoflagellate *Ceratium* divides during darkness. The day length may also control the initiation of spore formation rather than vegetative growth (e.g. in *Stephanopyxis*). Motile phototrophs, and heterotrophic protozoa with mutualistic algal symbionts such as *Prorodon*, may show phototaxis and thus be able to control, to some extent, the intensity of light to which they are exposed. The reaction of micro-organisms to light is not therefore a simple dependence on it for photosynthesis: the effects are complicated and wide-ranging, but poorly understood in most cases.

Currents

Currents are most obvious in rivers, but all large water masses have them and they are important for two main reasons. Firstly there is the problem of being washed away from a suitable habitat, though currents may also be used for beneficial dispersal. Secondly currents in the oceans, and to a lesser extent in large lakes, are responsible for transporting nutrients. Moving water also prevents the formation of nutrient deficient diffusion shells around sedentary microbes and this may be one of the reasons for the increased nutrient uptake in moving rather than still water. Currents in the surface waters may be separated from those below by the thermocline, indeed deep water currents may move in the opposite direction to those on the surface and usually they are slower (Fig. 7.1).

Nutrients

One of the major determinants of microbial biomass and variety in water is the supply of nutrients, both organic and inorganic. The amounts and chemical nature of dissolved and particulate organic matter vary enormously (Chapter 3, page 22), from virtually zero in springs to 1 or 2 mg/l (=ppm) in the deep oceans and oligotrophic lakes, and to high levels in eutrophic situations. In rivers and small shallow lakes most of the organic matter is allochthonous. While in large lakes and the sea most of the material is autochthonous.

Surfaces, which concentrate nutrients and micro-organisms by adsorption, may be of great importance at low nutrient levels: the water-air interface is a special surface, complicated by surface tension effects. Invertebrates comminute organic matter, so as to produce new surfaces for colonization which increases carbon turnover rates and stimulates microbial activity by the production of nutrient-rich faeces.

The available organic matter is important for heterotrophs, the low level accounting for the relative scarcity of bacteria in such habitats as springs, oligotrophic lakes and deep marine waters. High levels of dissolved organic matter may inhibit photosynthesis, though algae are normally the main primary producers in water and not only do they contribute their bodies to the food chains, via grazing or death; they also produce organic exudates (Chapter 3, page 22). This exudation sometimes leads to heavy growths of heterotrophs on the surfaces of autotrophs (see below, epiphytic communities). The interpretation of nutrient effects on living surfaces is complicated by the adsorption already mentioned, the release of oxygen during photosynthesis and by the release of antibiotic substances by some algae.

Inorganic nutrients, particularly nitrogen levels in the sea and phosphorus in fresh water (Chapters 4 and 5), may limit algal growth. The problems of nutrient availability in stratified waters have been mentioned above.

Pressure

Pressure, which increases by about 1 atmosphere for every 10 m depth, is greater than 300 atm in the majority of oceans and can reach 1200 atm in the greatest deeps. It has widespread effects on the physical and chemical properties of water and organisms, changing the pH, degree of ionization, viscosity, rate of membrane transport, solubility of nutrients, enzyme kinetics, molecular volume and the tertiary and quaternary structure of proteins. This complex of effects may be expressed in such easily observable phenomena as a change in cell morphology e.g. a change from a rod to filaments in some bacteria (further reading 5). Many organisms can stand short periods of high pressure, but death results even for barophilic microbes after prolonged periods at greater than 1500 atm. There is an interaction with temperature; higher temperatures result in greater pressure tolerance, but in natural environments temperature is low at great depths and indeed many of the deep ocean micro-organisms are both barophilic and psycrophilic. Some are obligate barophils so it is important to incubate cultures under pressure if counts are to be made (Table 7.1). Microbes are found even at the greatest depths in the ocean sediments, though fungi are rare below 4000 m and if there are live algae they must be living heterotrophically.

Table 7.1 The numbers per g wet weight of different groups of bacteria in the sediments at 7000 and 10000 m in the Pacific and Indian Oceans: plates incubated at different pressures. The populations contain considerable numbers of obligate barophils, and their optimum pressure appears to be close to that at which they were growing. (From C.E. Zobell and R.Y. Morita (1957) Barophilic bacteria in some deep sea sediments. *Journal of Bacteriology* **73**, 563–568.)

Location	Phillipine Trench; mean of 4 samples		Indian Ocean; mean of 2 samples		
Pressure at sample depth. Atmospheres.	1012–1021		702–725		
Incubation pressure Atmospheres	1	1000	1	700	1000
Total aerobes	5,500	550,000	750,000	1,670,000	10,900
Total anaerobes	5,500	100,000	—	—	—
Starch hydrolysers	780	550	55	1,000	0
Nitrate reducers	330	78,000	5,500	10,000	10
Ammonifiers	1,000	78,000	—	—	—
Sulphate reducers	0	52	0	55	0

— = no data available.

COMMUNITIES IN WATER

There are many distinct habitats in water, each with its characteristic community of micro-organisms. Firstly there are the forms which are maintained in the water column by turbulence, the plankton (Fig. 7.3). Secondly micro-organisms colonize surfaces, forming recognizably different communities, which are collectively known as benthos, on the different substrata: those growing on the surfaces of other plants (both angiosperms and other algae) are called epiphytic, those on animals epizoic, epipelic on mud, epilithic on rock and epipsammic in or on sand (Fig. 7.3). Thirdly there is a specialized community called the neuston attached to the air-water interface which occurs in the oceans as well as in the relatively still waters of lakes and ponds. This emphasizes yet again the importance of surfaces to micro-organisms. Though the species and genera may differ, there are similar, distinct communities of planktonic and benthic organisms all over the world, in both fresh water and the sea.

The different communities are not equally important in different waters: the deeper the water the less important the benthos in primary production. In most marine situations and in many lakes the majority of the benthos is well below the photic zone and so consists entirely of heterotrophs and chemotrophs, and in such deep waters the plankton near the surface is the only primary producer.

Plankton

The plankton consists of those organisms which normally exist for at least part of their life free in the open water: it should not include stray organisms washed in from the surrounding land or from other aquatic communities. It is usually divided into the phytoplankton (algae, fungi and bacteria) and the zooplankton containing protozoa and also a great variety of larvae and adults of copepods, crustacea etc., with which we will not be concerned except in their effect on algal populations by grazing. At the micro-organism level very similar and obviously closely related forms occur in the phyto- and zoo-plankton, particularly with the very small plankton (the nanoplankton). This part of the plankton is composed mostly of flagellates of the Haptophyta (mostly marine) and Chrysophyta (mostly freshwater) containing both pigmented and colourless forms which are 'claimed' by both the phycologists and the zoologists. The nanoplankton is often very important in the productivity of lakes and oceans but, since it passes through plankton nets and the fragile cells are damaged by all but the gentlest filtration, it has been little studied compared with the larger, more easily observed forms.

The main algae in the temperate phytoplankton are diatoms and dinoflagellates. In the sea some of the common diatoms are *Coscinodiscus*, *Chætoceros*, *Sceletonema* and *Thalassiosira*, while *Ceratium*, *Dinophysis*, *Exuviaella*, *Gymnodinium* and *Peridinium* are common dinoflagellates. Blue-green algae are not common planktonic organisms in the sea, though one genus, *Trichodesmium*, is regularly reported from the tropics, together with diatoms and coccolithophorids (Haptophyta). Dinoflagellates are not so common as in temperate oceans.

Fresh waters are more variable with diatoms such as *Fragilaria* and *Melosira varians* dominating neutral and slightly alkaline waters, together with species of *Ceratium* and *Peridinium* (Fig. 7.3). In more acidic, oligotrophic waters, in moorland pools for example, the desmids (*Cosmarium* and *Staurastrum*) are usually dominant though some diatoms (*Melosira distans*) and flagellates (*Chlamydomonas* spp. and *Euglena* spp.) may occur. Eutrophic waters tend to have more of the larger species while the oligotrophic lakes have small flagellates (e.g. Chrysophyta) as important parts of their flora. Where the organic matter levels are high there may be a preponderance of Euglenophyta (*Euglena* and *Phacus*) and Chlorococcales (*Scenedesmus*, *Chlorella* etc.). Cyanophyta may also be important components of eutrophic, fresh water plankton; there are both coccoid (*Microcystis* and *Coelosphaerium*) and filamentous forms (*Anabaena*, *Lyngbya* and *Oscillatoria*) though some of these may be colonies that have produced gas and risen from the sediments in or on which they normally live. Tropical lakes are characterized by larger species of plankton and also by genera, such as *Nitzschia*, which are benthic under temperate conditions. Eutrophic tropical lakes again have Cyanophyta as a major component. In arctic and alpine lakes the winter flora is dominated by small flagellates (*Chromulina*, *Mallomonas*, *Dinobryon* etc.), and these remain in the water during the brief ice free summer period.

In rivers, except in the slowest flowing parts, there is no true algal plankton, though there may be a considerable number of organisms drifting with the current which have originated in lakes, still backwaters and ponds which feed the river. Most of these rapidly disappear from the water downstream of their source, but some persist long enough to grow and divide in the river and so may be regarded as true plankton. Where true plankton does exist in slow rivers it consists of small centric diatoms (*Stephanodiscus hantzschii* and *Cyclotella*), flagellates and coccoid green algae (*Ankistrodesmus*, *Chlamydomonas*, *Pandorina*, *Scenedesmus* etc.).

Planktonic algae are not so important as the benthic flora and littoral communities in the productivity of estuaries; this is because of the extent of shallow water and also the suspended matter greatly reduces light penetration in the deeper regions. The population is dominated by marine diatoms though there are some small dinoflagellates (*Gymnodinium* and *Gonyaulax*). Coastal waters may however have higher numbers of planktonic heterotrophs than either the open ocean or the inflowing rivers. This is probably a reflection of higher nutrient levels because of run-off from the land or the suspension of sediments rich in organic matter. Pollution, especially from sewage, may also greatly increase the numbers (Table 7.2).

Protozoa are not usually a numerically important part of the plankton, except under bloom conditions. They feed on detritus, bacteria and the smaller algae and may also compete for soluble nutrients. In fresh water the population is dominated by ciliates (e.g. *Colpidium, Euplotes*) and colourless dinoflagellates; the Sarcodina are represented by heliozoa and testate amoebae (*Arcella* and *Difflugia*) which occur especially in polluted waters. Tintinnid ciliates (which secrete a shell or lorica around themselves e.g. *Codonella*) are also found in fresh water but they are much more frequent in the sea. Radiolarians are exclusively deep water marine organisms and marine foraminiferans (e.g. *Globigerina*) are found particularly in the tropics. Radiolarians and foraminiferans may have symbiotic, presumably mutualistic, dinoflagellates in them and there are also reports of foraminiferans with Chlorophyta and even diatom endosymbionts.

Planktonic bacteria are mostly Gram negative (80 to 95%) and often they

Figure 7.3. Examples of the genera of algae, and their growth forms, from the different communities in both fresh and salt water. Not to scale. (From F.E. Round (1977). *Algal Ecology*. Arnold, London. In press.)

KEY

Cy: Cyanophyta; Ch: Chlorophyta; X: Xanthophyta; B: Bacillariophyta; D: Dinophyta; Chry: Chrysophyta; R: Rhodophyta; P: Phaeophyta; E: Euglenophyta.

A. FRESHWATER

Epilithic: 1. *Calothrix* (Cy); 2. *Ulothrix* (Ch); 3. *Chaetophora* (Ch); 4. *Chamaesiphon* (Cy); 5. *Cymbella* (in tubes) (B).

Epiphytic: 6. *Oedogonium* (Ch); 7. *Ophiocytium* (X); 8. *Characium* (Ch); 9. *Tabellaria* (B); 10. *Cocconeis* (B); 11. *Gomphonema* (B); 12. *Dermocarpa* (Cy).

Planktonic: 13. *Pandorina* (Ch); 14. *Ceratium* (D); 15. *Mallomonas* (Chry); 16. *Fragilaria* (B); 17. *Melosira* (B); 18. *Staurastrum* (Ch); 19. *Anabaena* (Cy); 20. *Ankistrodesmus* (Ch); 21. *Chlamydomonas* (Ch); 22. *Coscinodiscus* (B); 23. *Dinobryon* (Chry).

Epipsammic: 24. *Navicula* (B); 25. *Chlorella* (Ch); 26. *Nitzschia* (B); 27. *Opephora* (B); 28. *Opephora* (B); 29. *Amphora* (B); 30. *Achnanthes* (B); 31. *Achnanthes* (B); 32. *Nitzschia* (B); 33. *Amphora* (B); 34. *Gomphonema* (B).

Epipelic: 35. *Caloneis* (B); 36. *Pinnularia* (B); 37. *Surirella* (B); 38. *Oscillatoria* (Cy); 39. *Spirulina* (Cy); 40. *Euglena* (E); 41. *Merismopedia* (Cy); 42. *Lyngbya* (Cy); 43. *Closterium* (Ch); 44. *Navicula* (B); 45. *Amphora* (B); 46. *Trachelomonas* (E); 47. *Cymatopleura* (B).

are motile, pleomorphic and pigmented. The classification of aquatic bacteria is difficult, they are usually assigned to the pseudomonads, the Corynebacteriaceae and Micrococcaceae (e.g. *Arthrobacter, Flavobacterium, Micrococcus, Pseudomonas, Sarcina, Spirillum* and *Vibrio*). Spore formers (*Bacillus* and *Clostridium*) become more important in nutrient-rich waters. When there is sewage contamination *Escherichia coli, Clostridium perfringens, Proteus vulgaris* and *Streptococcus faecalis* may be found.

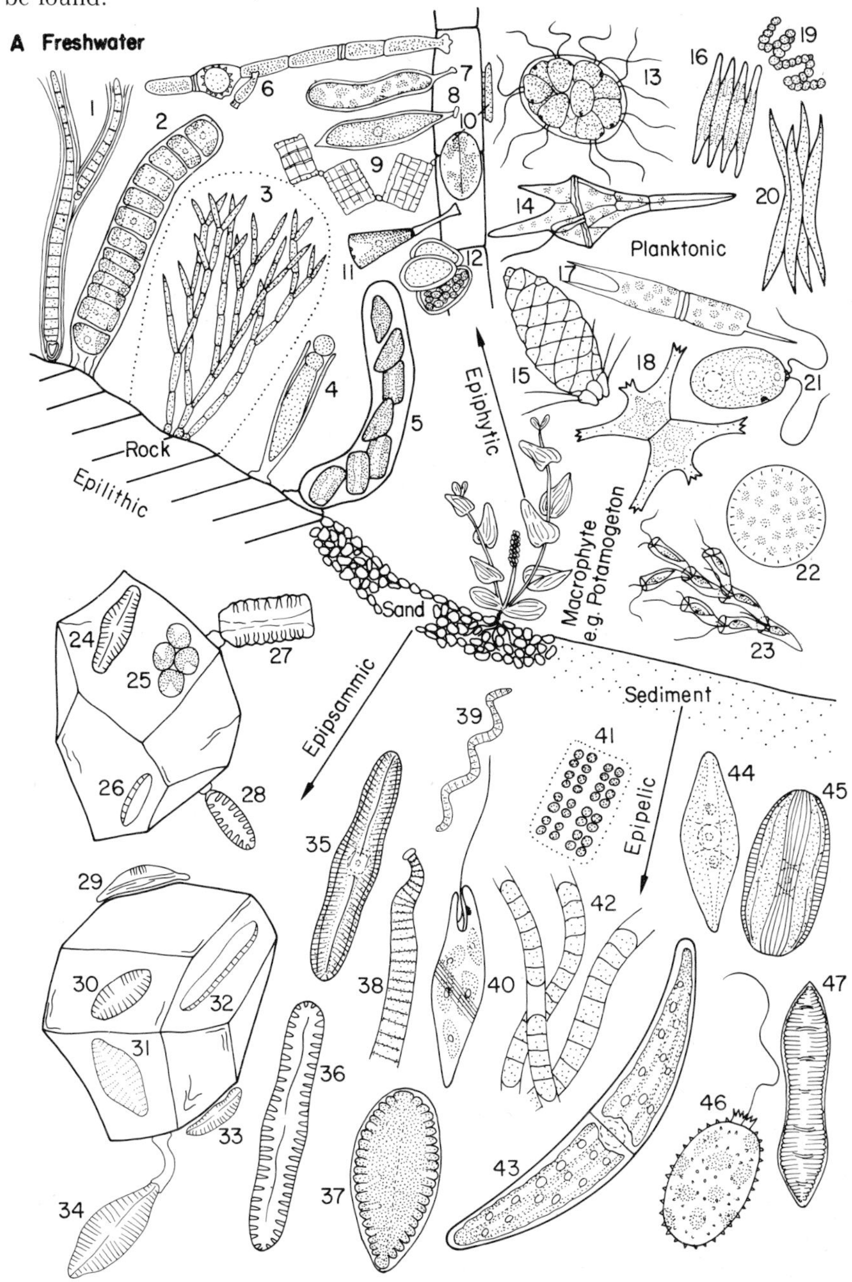

Table 7.2 Planktonic bacteria in the Bay of Naples, showing the great rise in numbers in the polluted inshore waters and the much smaller ones in the open sea. (From H.W. Jannasch (1955) Zur Ökologie der zymogenen planktischen Bacterienflora natürlicher Gewässer. *Archives für Mikrobiologie* **23**, 146–180.)

	Harbour	Bay of Naples	Open Sea
Total bacteria; numbers/ml	2.27×10^7	1.01×10^6	290
Saphrophytes; numbers/ml	3.64×10^6	2.10×10^5	82

The majority of the bacteria are not free in the water but are attached to particulate matter and their biomass is small compared with the primary producers, but their activity is most important in recycling nutrients, especially when the photic zone is isolated from the underlying water by a thermocline. Much of the phosphorus mineralization takes place in the epilimnion, often very rapidly with turnover times of only a few minutes in optimum conditions. Once material has sunk below the thermocline it must await breakdown of stratification before it can be re-used by most of the phytoplankton, so the amount of recycling in the photic zone is of great importance to the algal productivity.

The numbers of bacteria vary greatly, mostly in relation to the amount of organic matter, which may be either allochthonous or autochthonous. Viable counts commonly range from less than 100/ml to 1000/ml in lakes and down to only 1/ml or less in the open ocean, though total counts may give figures several orders of magnitude higher (Table 7.3).

Bacteriophages are known to occur in water, attacking bacteria, actinomycetes and Cyanophyta, though they are thought to be of little significance in controlling the populations. *Bdellovibrio* (page 25) has also been reported. Viruses that attack algae, fungi and protozoa are found in both fresh and sea water and animal viruses may be present if the water has been contaminated by faeces. Animal viruses do not apparently survive very long and transmission of viral diseases by water is comparatively rare, with the possible exception of infective hepatitis. Fish suffer from some virus diseases so the infective particles are possibly present in the water.

Fungi are not important planktonic organisms. In fresh water their spores,

Figure 7.3 continued

B. MARINE

Planktonic: 48. *Peridinium* (D); 49. *Rhizosolenia* (B); 50. *Hymenomonas* (Chry); 51. *Ceratium* (D); 52. *Coscinodiscus* (B); 53. *Chaetoceros* (B); 54. *Eucampia* (B); 55. *Thalassionema* (B); 56. *Sceletonema* (B).

Epiphytic: 57. *Erythrotrichia* (R); 58. *Pleurocapsa* (Cy); 59. *Cocconeis* (B); 60. *Grammatophora* (B); 61. *Bulbochaete* (Ch); 62. *Synedra* (B); 63. *Licmophora* (B); 64. *Cymbella* (B).

Epilithic: 65. *Fragilaria* (B); 66. *Navicula* (in tubes) (B); 67. *Rivularia* (Cy); 68. *Ectocarpus* (P); 69. *Bangia* (R).

Epipelic: 70. *Amphidinium* (D); 71. *Amphora* (B); 72. *Holopedia* (Cy); 73. *Hantzschia* (B); 74. *Mastogloia* (B); 75. *Pleurosigma* (B); 76. *Nitzschia* (B); 77. *Diploneis* (B); 78. *Amphiprora* (B); 79. *Navicula* (B); 80. *Tropidoneis* (B).

Epipsammic: 81. *Licmophora* (B); 82. *Coscinodiscus* (B); 83. *Cocconeis* (B); 84. *Opephora* (B); 85. *Cocconeis* (B); 86. *Licmophora* (B); 87. *Raphoneis* (B); 88. *Amphora* (B).

originating in the benthos, are dispersed in the water but they do not grow there. There are epiphytic and endophytic parasites of algae, mostly Chytridiales and Saprolegniales (page 121), but the only free living planktonic fungi are the yeasts and these do not occur in any great numbers except near shore lines where they are soil forms. There are some stenohaline (page 96), obligate halophils amongst the marine yeasts (e.g. *Metschnikowia*) but euryhaline forms are more common

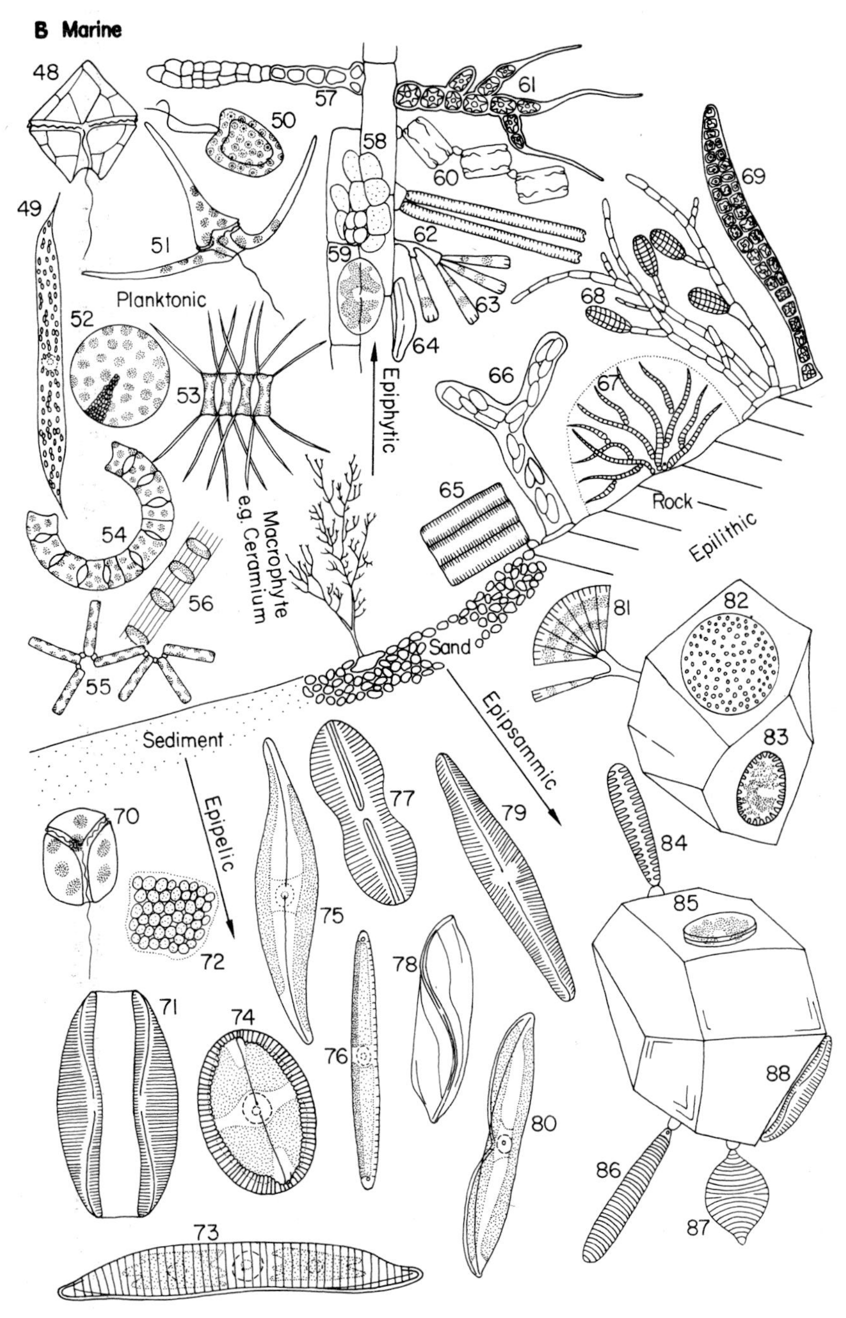

Table 7.3 The numbers of bacteria in the plankton of different sorts of fresh water, with very low numbers in springs and oligotrophic lakes but much higher numbers in eutrophic waters (shown by the low oxygen and the high nitrate concentrations) and near the bottom sediments of lakes. (From H.W. Jannasch (1955) Zur Ökologie der zymogenen planktischen Bacterienflora natürlicher Gewässer. *Archives für Mikrobiologie* **23**, 146–180.)

	River Aale			River Fulda			Lake Sieburger 3.5m deep and eutrophic		Lake Viewaldstätter 45m deep and oligotrophic	
	Spring	20m from spring	2000m from spring	Upper course	Middle course	Lower course	½m from surface	3m from surface	½m from surface	44m from surface
Total count numbers/ml	0	67,800	614,000	352,000	9.8×10^6	845,500	300	1.03×10^6	120*	56,700
Saprophytes numbers/ml	85	83,000	135,000	68,000	1.2×10^6	95,000	840	3.93×10^5	180	19,100
Oxygen mg/l	0.60	11.50	8.75	9.45	2.21	4.82	8.95	4.40	9.11	0.30
Nitrate mg/l	0	1.0	8.0	4.5	11.0	4.0	7.5	2.0	—	—

* mostly Aufwuchs (=epiphyton) not true plankton — not determined

(e.g. *Candida parasilopsis, Cryptococcus, Debaryomyces,* and the basidiomycetes *Leucosporidium, Rhodosporidium* and *Sporobolomyces*). The numbers in the open sea are very low (ranging from 1 to 500 cells/l, usually 10 to 40/l). The coastal rise in numbers is also found with filamentous fungi and *Cephalosporium, Cladosporium, Penicillium* etc. may be isolated but are probably not part of the true plankton. Benthic fungi may also be brought into suspension by wave action and turbulence around coasts and shores.

The distribution of plankton in any large body of water is not uniform: the numbers and species vary from place to place. Horizontally the patches of different plankton may be many kilometres wide in the oceans of a few metres or tens of metres in small lakes: the changes may occur over only a few metres depth (e.g. Fig. 5.3, *Chromatium*). The factors controlling the abundance of the phytoplankton include the nutrient status, temperature, light intensity and duration, and the grazing pressure exerted by the consumers: these vary with the seasons. Particularly outside the tropics there are therefore seasonal fluctuations in the numbers of most planktonic organisms, especially algae (Fig. 7.4), and upon them depend the heterotrophs whose population changes often follow, sometimes after a short lag, those of the primary producers (Fig. 7.7, page 109).

The simplest situation is a single peak in numbers during the summer which is initiated by the increased intensity and length of light in the spring; such a pattern may develop in polar regions (Fig. 7.4) where a thermocline does not usually occur. In the Antarctic diatoms are a major component of the flora in the summer 'bloom', but Arctic blooms are mainly caused by increase in the nanoplankton, mostly small flagellates such as Chrysophyta and Haptophyta. Dinoflagellates are important in all polar seas.

A single spring or early summer peak may occur in temperate regions but more usually a stable summer thermocline develops and limits the nutrients in

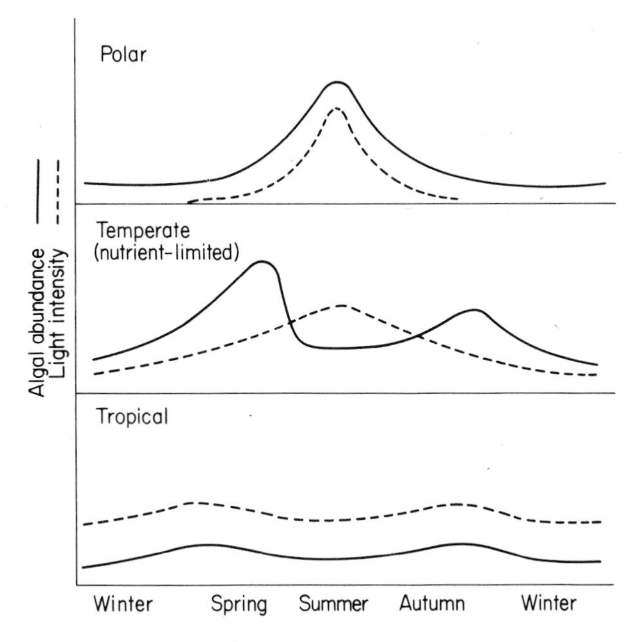

Figure 7.4. Diagram of the possible patterns of algal blooms in different climatic regions.

the epilimnion, or the marine surface waters, so that there is a decline in numbers (Fig. 7.4 and Fig. 5.6 for silica depletion) as nutrient deficiency develops. A smaller autumn bloom may occur as the nutrients are recycled and the thermocline breaks down. The limitation of the bloom may not be due solely to nutrient depletion; in autumn reduced photosynthesis caused by turbulent mixing which takes the algae intermittently below the photic zone, accentuated by shortening days and the lower light intensity, may play a part in reducing the bloom and the effects of grazing and parasites are also important (see below).

These fluctuations in numbers represent the sum of all the species present, though frequently blooms are almost monospecific. Thus the spring bloom in many temperate eutrophic lakes is of diatoms (*Asterionella, Fragilaria, Melosira, Stephanodiscus* or *Tabellaria*) and is often terminated by silica depletion (Fig. 5.6). Chlorophyta may also occur and, as the diatoms decrease, they may become dominant in the summer (*Ankistrodesmus, Chlorella, Cosmarium, Pediastrum* etc.). Cyanophyta such as *Anabaena flos-aquae, Chroococcus* and *Oscillatoria* may also occur in the summer and autumn and be replaced in their turn by Dinophyta (*Ceratium* and *Peridinium*). In the winter and early spring the smaller forms, like the Cryptophyta, predominate (*Rhodomonas* and *Cryptomonas*). This generalized pattern is obviously subject to differences between particular bodies of water, but some at least of these stages are found in most fresh waters in temperate regions (e.g. Fig. 7.5). In the sea the sequence is usually less complicated, the spring diatoms (*Sceletonema, Thalassiosira*) being replaced by dinoflagellates, though there may again be species successions within each peak; smaller species of diatoms usually occur first.

Under tropical conditions these seasonal variations are not so pronounced, though in the sea there may be a slight spring increase followed by a summer low due to depletion of nutrients (Fig. 7.4). Eutrophic tropical lakes may have almost continuous bloom conditions, particularly of Cyanophyta.

These successions are usually very consistent, with the same species appearing year after year in a particular lake or region of the sea and even the generalized sequences are remarkably constant within the same climatic zone. What determines the sequence is largely unknown. Changing light and nutrient status are presumably important and different species are known to have different optima. There is a suggestion, from marine and laboratory studies mostly, that the availability of vitamins may control the abundance of some algae. Vitamin B_{12} is often required and is produced by some bacteria and also by other algae e.g. *Coccolithus huxleyi*, which is itself dependent on an exogenous supply of thiamin (B_1) produced, together with biotin, by *Gonyaulax polyedra, Skeletonema costatum* and *Stephanopyxis turris* among many others. It is possible that the later organisms in a sequence are dependent on the earlier ones for vitamins, so determining the sequence. Algae are also known to produce antibiotics against other algae (and bacteria, page 121) and this could be a factor in keeping blooms to one or a few species.

The increased activity of some species is dependent on the mixing of the water, particularly in the late winter and early spring. These species (e.g. *Melosira italica* subsp. *subarctica* and *Gloeotrichia echinulata*) either form spores, or have vegetative cells with no buoyancy, in the littoral zone and on the sediment surface where they

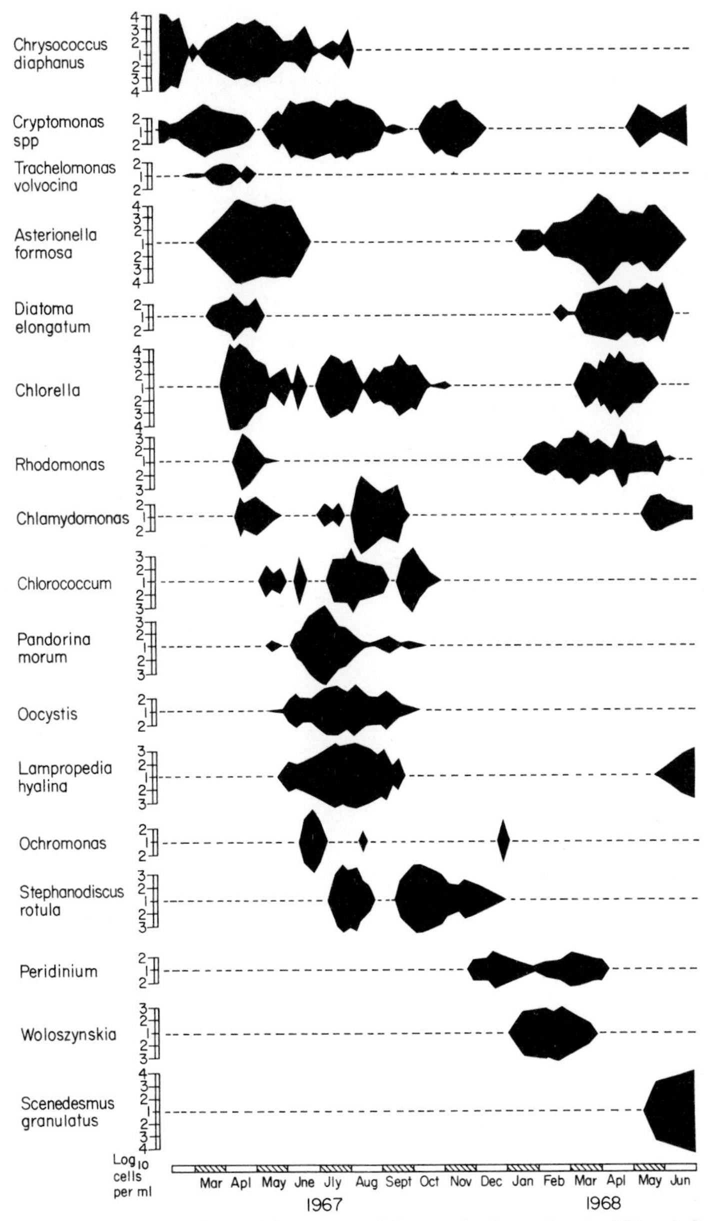

Figure 7.5. The species succession and frequency of algae and a bacterium in Abbott's Pool, a eutrophic pond near Bristol. (From F.E. Round (1971). The growth and succession of algal populations in freshwaters. *Mitteilungen Internationale Vereinigung für theoretische und angewandte Limnologie* **19**, 70–99. Based on data supplied by Dr C.M. Happey.)

remain until mixed into the main water mass by turbulence. Given enough light, they may then initiate a bloom. However this mechanism is not possible in the ocean, for most sediment is too deep for mixing into the photic zone, and there are many algae, both fresh water and marine, that are not known to form spores or other resting stages. These almost certainly exist in the water column at all times, though the very low numbers may make them almost impossible to find without sampling and concentrating very large volumes.

Wash-out by through-flowing water may influence the plankton inoculum especially in small lakes where the residence time is short. Seasonal rains obviously determine the period of wash-out or in cold climates the spring thaw may be the time of greatest flow. The effect of wash-out is most noticeable if it occurs at a time of low temperature or low light intensity when the reproduction rate is very slow.

Grazing by copepods and other zooplankton has been held responsible for the decline of marine plankton blooms. In freshwater protozoa and rotifers seem to be the most important grazers. Protozoa are also parasites: their numbers start to rise when the phytoplankton is well established and then the latter decreases rapidly as a high proportion of the cells are attacked (Fig. 7.6A). The predators or grazers are presumably present in the water all the time but the chances of them finding their prey, (or being seen to do so by the investigator!)

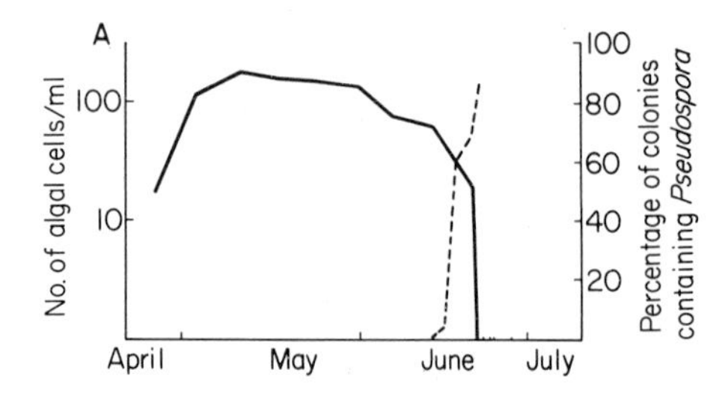

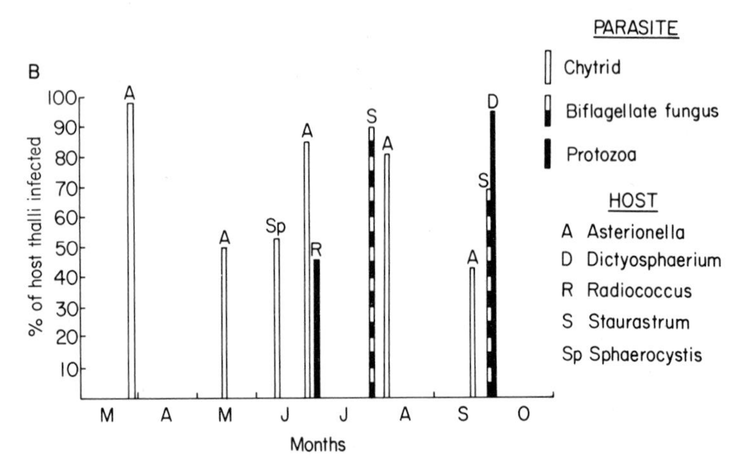

Figure 7.6. The effects of parasites on plankton blooms. A. Decrease in a bloom of *Gemellicystis imperfecta* (———) as the number of colonies infected with the protozoan *Pseudospora* increases (– – – –). (From H.M. Canter and J.W.G. Lund (1968). The importance of protozoa in controlling the abundance of planktonic algae in lakes. *Proceedings of the Linnean Society* **179**, 203–219.)
B. Ten epidemics of various parasites of plankton which resulted in a subsequent decrease of the algal numbers in Grasmere in the English Lake District during the summer of 1973. (From data in 42nd Annual Report, Freshwater Biological Association (1974). Data of Drs H.M. Canter and J.W.G. Lund.)

are small except in bloom conditions. Grazing may be a factor in maintaining the summer minima in the algal population of temperate waters and selective grazing may affect species competition and succession. Various fungal parasites (especially chytrids, see below page 121) have also been shown to limit fresh water blooms (Fig. 7.6B). Parasitism or grazing is not usually uniform: there are patches of intensive feeding which may almost eliminate the alga, while nearby there are ones which are unattacked. This accentuates, or in some cases may be the reason for, the uneven distribution of plankton already noted. In bloom conditions the zooplankton may exhibit 'superfluous feeding' in which the food is only partially digested before being voided and this rich faecal material is rapidly recycled by bacteria and also is an important source of benthic organic matter. The bacterial numbers rise during and after algal blooms (Fig. 7.7), either because of direct effects of exudation by the algae or because of this secondary faecal effect. Bacteria are also grazed, e.g. by protozoa, and the very low numbers in the open oceans have been attributed to grazing pressure as well as very low nutrients.

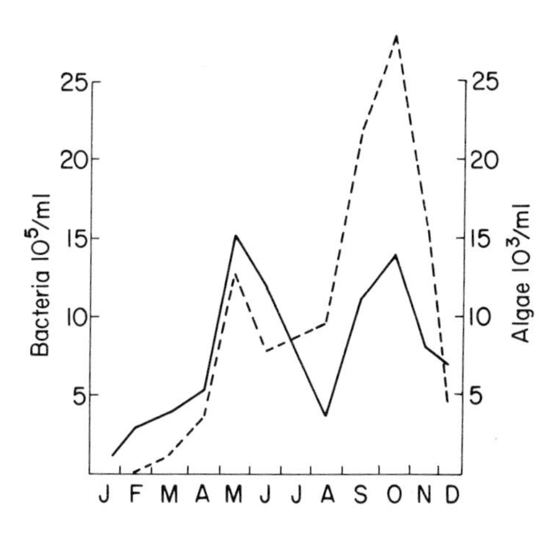

Figure 7.7. The simultaneous rise and fall in the numbers of heterotrophic bacteria (———) and the planktonic alga *Scenedesmus* (– – –) in an artificially enclosed portion of a lake. (From J. Overbeck and H.D. Babenzien (1964). Bakterien und Phytoplankton eines Kleingewässers im Jahreszyklus. *Zeitschrift für Allgemeine Mikrobiologie* **4**, 59–76.)

Some algae produce toxins which may reach levels sufficient to kill zooplankton and even fish, and this may be partly responsible for the exclusion of grazers from ungrazed regions of blooms. In some spectacular blooms, such as the 'red tides' caused by *Gymnodinium brevis* and *G. splendens* the toxins may reach such levels as to make seafood poisonous to man.

Maximum algal abundance lies below the surface, probably because photosynthesis is inhibited by strong light at the top of the water column. In the absence of a thermocline there is a fairly uniform distribution with depth because of the mixing, but with a stable thermocline there is a concentration of the microbial biomass in the surface water, with the maximum algal abundance between 5 and 100 m, varying with the particular water body considered. The bacterial maximum is usually at the same depth, or sometimes deeper, than the phytoplankton (Fig. 7.8) and may be on the thermocline if this is near the surface. Organisms and particulate organic matter may accumulate at any boundary where less dense water rests on a denser layer; the rate of sinking will be reduced or stopped at these boundaries.

Numbers of all aerobic organisms fall rapidly below the thermocline, though they may rise again near the bottom sediments (cf. oxygen distribution in Fig. 7.1). Anaerobic bacteria producing methane (*Pseudomonas methanica*), reduced sulphur compounds (*Desulfovibrio*, *Thiotrix*) or the photosynthetic bacteria (*Chlorobium*, *Chromatium*, *Rhodopseudomonas*) may have their maximum numbers in these deeper waters.

The abundance of yeasts also decreases with depth; though there is very little information they probably respond like the heterotrophic bacteria with a maximum near or below that of the algae. Parasitic fungi such as the chytrids are of course linked to the distribution of their hosts.

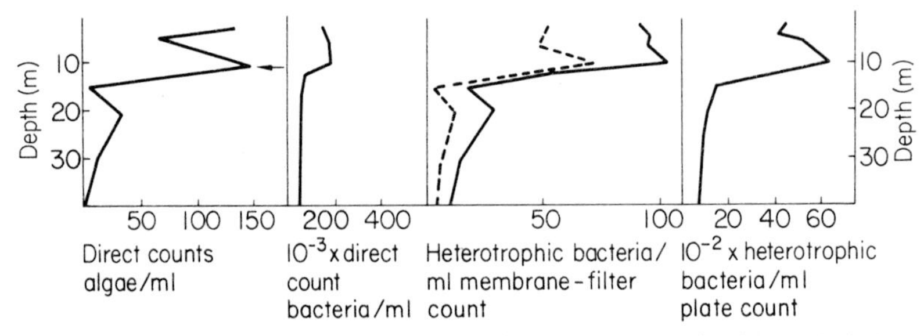

Figure 7.8. The variation in the numbers of algae (diatoms) and bacteria with depth during a bloom in Lake Windermere. The arrow in the left hand graph marks the position of the thermocline: for heterotrophic bacteria in membrane filter counts, --- is attached bacteria and ——— is total bacteria. (From J.G. Jones (1972). Studies on fresh water bacteria: association with algae and alkaline phosphatase activity. *Journal of Ecology* **60**, 59–75.)

The phytoplankton may move up or down in a diurnal cycle. How it does this is not known, for except in the calmest waters, the motility of the individual organisms, or the flotation mechanisms in non-motile forms like *Anabaena* or *Microcystis* is unlikely to account for much compared to the turbulent movement induced by the water. Usually there is an upward movement of the plankton during the day and a downward movement at night (e.g. for many dinoflagellates such as *Ceratium*). The pattern of movement does however vary and the reverse response is known, and some, such as *Peridinium bahamensis*, move to the surface twice each day. It has been suggested that these movements are concerned not only with seeking light in the upper layers but also with the occurrence of richer nutrients in the deeper water. Furthermore the movements may continue under laboratory conditions in the absence of any periodic stimulus, and so are presumably inbuilt rhythms, controlled by 'biological clocks'. There are also diurnal changes in physiological processes: photosynthesis usually has a peak rate at mid-day or shortly afterwards, and there may be a daily fluctuation in the chlorophyll level. Cell division and even bioluminescence (in *Gonyaulax polyedra*) have diurnal rhythms.

Apart from the movements of the algae there is the general problem of their buoyancy. Most plankton is denser than water and should sink, indeed it does in still water or when dead. However many live cells either sink more slowly than their size would suggest they should or they may even rise. The floating may be caused by gas vacuoles in the Cyanophyta, but the efficiency of oil droplets as floatation devices has been questioned. The ionic concentration of the vacuolar sap may also affect the density of the cells.

Neuston

This is the specialized community usually sampled with the plankton, if at all, which lives associated with the surface film. The habitat is unusual in many ways compared with the rest of the lake or the ocean; it is subject to rapid temperature fluctuations, high light intensity and is very well aerated. The neuston has been little investigated but seems to be of universal occurrence, even if the waters are rough and the surface broken as in most of the ocean. Organic matter, both particulate and dissolved, and inorganic nutrients such as phosphate, accumulate on the surface of the film and there are great increases (up to 300 fold over numbers in the top ten metres of water) in bacteria such as *Pseudomonas* and *Caulobacter*. Flagellates, particularly the small Chlorophyta and Chrysophyta that would be assessed as nanoplankton in most studies, also show very great increases, and ciliates lesser ones. There are reports from fresh waters of specialized amoebae (*Arcella* sp.) living on or under the water surface.

Epilithon and related communities

Epilithic communities (Fig. 7.3) vary in their importance and species composition in different habitats, though they share common morphological and physiological characteristics imposed upon them by their environment. The epilithon is insignificant below the photic zone; most of the bottom deposits are sediments rather than rocks, with sands and gravels in lakes and organic muds in the oceans. However in the littoral zone the epilithon may be of great importance and have a high productivity, though the habitat is very specialized in marine situations because of the tidal rhythm.

In springs and fast-flowing rivers the epilithon contains the main primary producers and also many organisms belonging to other trophic levels. The movement of water over the surface is a major problem, and most micro-organisms are encrusting or basally attached though in very fast flowing waters even this may not be sufficient for the movement of the stones may prevent the establishment of a permanent community e.g. in the rapids of rivers. The position on the rock, in relation to the current, affects the population and there are characteristic algae facing the current e.g. *Lithoderma fluviatilis*, *Cocconeis placentula*, *Ulothrix* and the stalked diatom *Gomphonema*. *C. placentula* also occurs in the turbulent zone on the lee side of rocks. A few lichens, particularly of the genus *Verrucaria*, also regularly occur on submerged rocks in temperate regions. Probably because of the exacting environment there are often almost 'pure stands' of one alga. There are characteristic attached bacteria such as the filamentous *Sphaerotilus* and the stalked *Caulobacter* and *Hyphomicrobium*. These may be using nutrients in the flowing water, or those from the algae or adsorbed onto the rock surface. Generally low organic nutrient levels result in very poor growth except in association with algae. Experimental colonization of stones and artificial substrates in water suggests that a general bacterial flora (*Pseudomonas, Flavobacterium* and *Achromobacter*) establishes itself first as a slime layer, followed by *Caulobacter* and *Hyphomicrobium*, and only then do diatoms and attached protozoa and those with thigmotactic cilia (respond to contact) become established. Later the macroscopic algae may be important: *Cladophora glomerata* is common in temperate regions in fresh water.

There are often seasonal changes in the epilithic river and lake flora, similar to those in the plankton, with a spring diatom bloom followed by Chlorophyta in the summer and sometimes a blue-green algal peak between these two. Nutrients are not usually limiting in the epilithon on lake shores which are supplied not only from the lake water, but also from the littoral sediments and run-off from the land.

Stones on the lake shore are dried when the water level fluctuates but such stones are always covered with diatoms, blue-green algae, desmids and bacteria, usually all embedded in mucilage. There may even be some zonation of species on lake shores, similar to that on sea costs though diatoms have a fairly uniform distribution. The dominant organisms in the upper regions, subject to the greatest desiccation, are the Cyanophyta *Gloeocapsa* and *Scytonema* (cf. terrestrial rocks, page 74) then moving nearer the water *Nostoc* and *Calothrix*, then *Tolypothrix*, with *Rivularia*, *Schizothrix* and the Chlorophyta *Spirogyra* and *Oedogonium* in permanently submerged conditions.

The marine epilithon is dominated by the brown algae and other macroscopic forms whose zonation down the shore has been so much studied. Microscopic algae are however present and may also show zonation. The rocky sea shore is very inhospitable with repeated desiccation and wetting, and the associated changes in salinity, and rapid fluctuations in temperature. Furthermore there is very heavy grazing pressure by molluscs, arthropods and sometimes fish. The supratidal zone has particularly serious problems of high salinity from evaporating spray, which may well be followed by rain with a rapid reduction in salinity. Such an extreme habitat is only colonized by lichens (e.g. *Verrucaria* spp.) and Cyanophyta (*Calothrix scopulorum, Gloeocapsa* and *Lyngbya* spp.) which may fix nitrogen at a level (2.5 $gN/m^2/year$) which is significant in an environment so poor in organic nutrients. On very porous rocks like chalk cliffs there may be a more extensive supratidal flora including some diatoms in cracks and crevices and also communities dominated by Chrystophyta. There are also considerable quantities of detritus in the supratidal zone, mostly composed of seaweed which is decayed by bacteria (*Cytophaga*, *Vibrio*) and marine ascomycetes (*Helicoma, Halosphaeria*). Guano is a special form of supratidal detritus and, apart from the heterotrophs, may be colonized by *Prasiola* (page 69).

There is very little information on the micro-organisms within the intertidal and subtidal zones, but the slime layers on rocks are composed of bacteria, diatoms, coccoid Chlorophyta and blue-green algae (e.g. *Rivularia mesenterica*). As in freshwater the initial colonization is by bacteria and microscopic algae. The recent example of the colonization of the island of Surtsey (see also page 75) showed that the first arrivals were bacteria and *Navicula* and *Nitzschia* spp.

Coral reefs are a specialized and highly productive marine environment in which there is extensive epilithic growth in the more exposed regions, particularly of *Porolithon* (Rhodophyta) which is important in depositing calcium carbonate. Other red algae cement reefs together and on the windward side may take the main force of the wave action. Diatoms commonly grow on the coral blocks in deep water off the reef edge and in the lagoons, where blue-green algae (e.g. *Calothrix*, *Nostoc* and *Lyngbya*) are also important. The latter occur on the rocky and sandy beaches as well. Since there is no nitrogen available from land drainage on atolls the nitrogen fixation by these Cyanophyta could be very important in the productivity of reefs. Very high values of fixation have been recorded, for example nearly 2 kgN/ha/day has been measured on intertidal reef flats where the main organism was *Calothrix*. This compares very favourably with the fixation rates on land (cf. Table 4.1). In addition there are blue-green algae living endolithically within the coral skeleton and Cyanophyta are also symbiotic in some sponges. Algal symbionts also recycle nitrogen in the coral reef. The most important symbionts on reefs are dinoflagellates (e.g. *Gymnodinium microadriaticum*) which live within the coral polyps and are responsible for a considerable proportion of the photosynthesis on the reef. Symbiotic dinoflagellates are a very important feature of most tropical seas (see page 100).

Epipsammon and associated communities

Growth will only occur on sand grains if they are relatively undisturbed by currents and wave action, and even then the micro-organisms are usually small colonies in crevices or on concave surfaces where they are somewhat protected from abrasion. Such sites usually have a layer of organic matter, either adsorbed or produced as mucilage by the organisms. The flora (Fig. 7.3) consists of bacteria and small diatoms, either encrusting or stalked, (e.g. *Achnanthes*, *Amphora*, *Fragilaria*, *Opephora* in fresh water and *Cocconeis*, *Opephora* and *Raphoneis* in the sea). In addition there is usually a distinct community of motile micro-organisms, e.g. diatoms (*Nitzschia*, *Hantzschia*) and blue-green algae. Flagellates and euglenoids may occur if nutrients are more abundant, by the mixing of organic mud with the sand for example. Ciliates are the most common protozoa and their morphology and physiology show adaptation to their habitat; species with long thin bodies and pronounced positive thigmotaxis being dominant (*Remanella*, *Spirostomum* and *Trachelocerca*).

In marine littoral sands *Hantzschia* migrates to the surface as the tide recedes and moves downwards again as the tide comes in: this is probably a mechanism to prevent wash-out and abrasion damage in the disturbed surface layers. A very similar rhythm is shown by the flat worm *Convoluta roscoffensis* containing endosymbiotic algae (*Platymonas convolutae*) which live in some intertidal sands especially on the northern coasts of Brittany. The host animal ceases to feed on reaching maturity, relying on the symbiont not only for the products of photosynthesis, but also for the conversion of nitrogenous waste products back into usable amino acids.

Epipsammon, in fresh water, may show a seasonal rhythm with maximum numbers and productivity in spring (Fig. 7.9), just as in plankton. The cause of the spring bloom does not however seem to be light intensity for phytoplankton also increases at this time and shades the sand surface, and as light intensity increases at the end of the plankton bloom the epipsammon decreases.

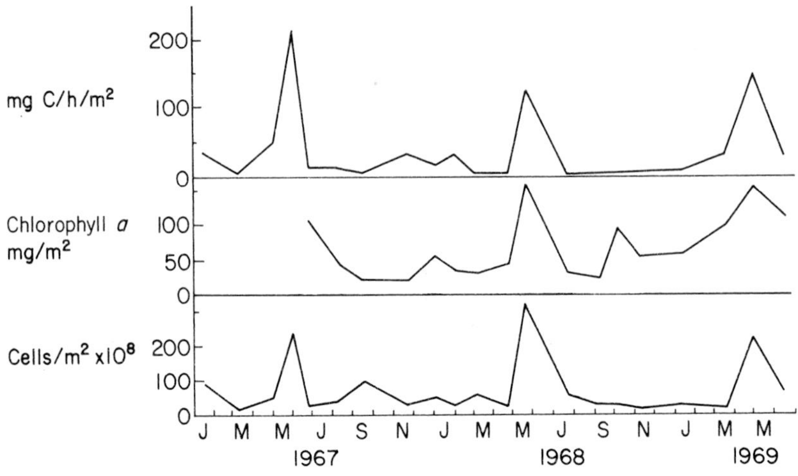

Figure 7.9. The numbers, chlorophyll a content and potential carbon fixation of a lake epipsammic community consisting mostly of diatoms. (From M. Hickman and F.E. Round (1970). Primary production and standing crops of epipsammic and epipelic algae. *British Phycological Journal* 5, 247–255.)

Epipelon and associated communities

There is tremendous variation in muds, from the organic ones in the ocean deeps, to salt marshes and the inorganic river silts and lake bottoms. There is a corresponding diversity in the microbial communities which inhabit them (Fig. 7.3). Microbial numbers and activity are usually greatest in shallow-water muds and algae are only active in the photic zone, and not therefore in the sediment that covers most of the world in the ocean deeps. There are again reports of heterotrophic growth, particularly of diatoms.

In freshwater muds the main factor controlling the species present is the water type, particularly the acidity and nutrient status. In peat bogs desmids may be dominant but all other muds have diatoms as the major group of algae with varying numbers of Chlorophyta, Cyanophyta and Euglenophyta. Some idea of the relative proportions of the various epipelic organisms can be gained from their respiration rates: in a temperate lake, in June, bacteria accounted for 31% of the respiration, algae for 23%. protozoa for 13% and the total benthic metazoa 33%.

The main protozoa are ciliates, usually heterotrophs living by ingesting detritus and other microbes. There are some mutualistic associations, where algae occur as endosymbionts. Such associations are known for several protozoa such as *Frontonia, Prorodon, Euplotes* and *Stentor* but the most studied one is *Paramecium bursaria*. The alga is invariably *Chlorella*. These associations are probably not at all important in the productivity of most fresh waters (cf. tropical marine conditions, page 113), but they are of scientific interest.

Springs are specialized habitats where the main community is epipelic (occasionally epilithic). The environmental conditions are remarkably uniform because of the underground source of the water; the temperature is contant, usually 9 or 10°C in temperate regions, the water is low in nutrients especially organic materials, but saturated or supersaturated with bicarbonate, particularly in limestone districts. Carbonate may be deposited as the excess carbon dioxide equilibrates with the atmosphere or is removed by photosynthesis. Blue-green algae in particular are associated with crusts of precipitated carbonates. The microflora is sparse but otherwise typical of low nutrient waters. Macroscopic algae e.g. *Batrachospermum*, may occur and they are important as a nutrient source for their microscopic epiphytes. Iron bacteria (*Gallionella*, *Leptothrix* and *Crenothrix*) are often present, since some of them are chemotrophs, and similarly sulphur springs often have sulphur oxidizing bacteria, though these are usually also thermal springs and so have a specialized flora e.g. *Leptothrix thermalis* (iron) and *Chloroflexus* (sulphur). Hot springs frequently have a high mineral content and are very exacting environments in which Cyanophyta dominate, particularly at temperatures between 50 and 60°C (e.g. *Mastigocladus laminosus, Synechococcus lividus*). At lower temperatures diatoms such as *Achnanthes* may occur, and there may be abundant growth of bacteria amongst the algae, utilizing the exuded organic matter, including nitrogen fixed by the blue green algae. High temperature strains of protozoa are also found including *Actinophrys sol, Cyclidium glaucoma, Hyalodiscus limax* and *Nassula elegans*.

The algae in the epipelon of rivers are more varied than in springs, though again dominated by motile diatoms in temperate regions, with flagellates such as *Phacus* and *Euglena* being present when there is a high concentration of organic matter.

The epipelon of oligotrophic lakes consists of desmids and such diatoms as *Frustulia rhomboides, Eucocconeis flexella* and *Eunotia* and filamentous Chlorophyta (*Spirogyra* and *Zygnema*). Eutrophic waters have more Cyanophyta (*Aphanothece* and *Nostoc*) along with Chlorophyta and euglenoids. In arctic lakes the epipelon may be the source of a major part of the total productivity, for plankton is sparse especially early in the season: green algae are usually dominant though diatoms are still present in high numbers (60 cells/mm^2).

The epipelic algae show seasonal fluctuations in numbers, usually with only one main bloom (Fig. 7.10): the two examples in Figure 7.10 have the peak numbers at different times of the year, one is in mid-winter and mostly of diatoms, the other is of blue-green algae with a summer bloom, as in the plankton. The reason for the collapse of the bloom is not known; nutrients which are important in the plankton are rarely limiting in mud, and as in the epipsammon, light would not be a major factor. There may also be a diurnal variation in the epipelon numbers at the surface as motile forms come up during the day: this is apparently a phototaxic response for it does not persist in continuous light (cf. page 110).

Bacterial numbers in fresh water muds are of the order 10^4 to 10^6 per g dry weight near the surface though the numbers, even of anaerobes, fall rapidly with greater depth into the sediment (Fig. 7.11). Actinomycetes occur particularly in eutrophic lakes and in the presence of high levels of organic matter, though the numbers are low (e.g. < 10/ml sediment). Numbers in rivers may be much higher (1000's/ml). The genera which produce dry spores are common (*Micromonospora, Nocardia, Streptomyces*) but there are also less well known genera (e.g. *Streptosporangium*) and some with motile spores (e.g. *Actinoplanes*) which are truly aquatic.

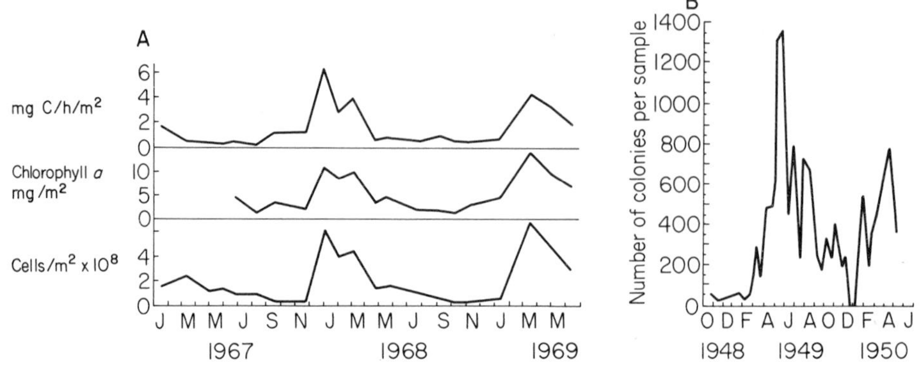

Figure 7.10. A. The seasonal cycle of growth of epipelon in a lake where the most important organisms were diatoms: the peak population and productivity is in the winter. (From M. Hickman and F.E. Round 1970. Primary production and standing crops of epipsammic and epipelic algae. *British Phycological Journal* 5, 247–255.) B. Seasonal cycle in the epipelic blue-green algae in Lake Windermere. The peak occurs much later, in May or June. (From F.E. Round, 1961. Studies on bottom-living algae in some lakes in the English Lake District. V. The seasonal cycle of Cyanophyceae. *Journal of Ecology* **49**, 31–38.)

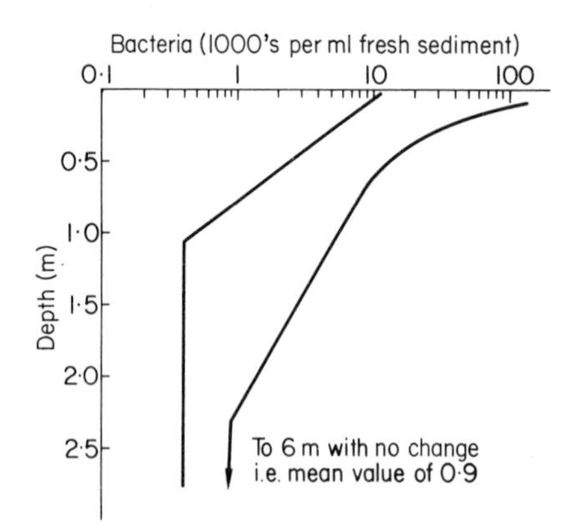

Figure 7.11. The variation in bacterial numbers (both aerobes and anaerobes) with depth in the sediment of two acid lakes. The distribution pattern is similar for both lakes, though the depth at which the numbers reach a steady, low level, varies. (From F.R. Hayes and E.H. Anthony (1959). Lake water and sediment VI. The standing crop of bacteria in lake sediments and its place in the classification of lakes. *Limnology and Oceanography* **4**, 299–315.)

Oxygen distribution has a great effect on fungi in muds and there may be very low numbers during the summer stratification when the hypolimnion is deoxygenated. *Achlya, Pythium* and *Saprolegnia* species are the most commonly reported, though there are also large numbers of deuteromycetes near the shore which are most probably contaminants from soil, since they are not adapted to sporulate and be dispersed in water. One important exception is the occurrence of aquatic hyphomycetes which grow extensively on leaves, twigs etc. in and on aerobic sediments and have specialized spores for water-borne distribution (further reading 4). Frequently the spores are branched or have appendages (Fig. 7.12) which may reduce the sinking rate but are probably most important in assisting the deposition of the spores (e.g. *Tetracladium, Clavariopsis* and *Varicosporium*). The spores are most abundant during the winter, though whether this reflects a spore production phase or a more general increase in activity at this time is not known. Yeasts with similar branched groups of cells (e.g. *Candida aquatica*) also occur and aquatic ascomycetes are common on debris, particularly wood and twigs, and again they have spores with appendages. There are successions of fungi on decaying organic matter in water, many of the hyphomycetes being early colonizers (e.g. *Clavariopsis aquatica*) along with *Pythium, Saprolegnia* and *Fusarium*. Other hyphomycetes and the ascomycetes invade the substrate later.

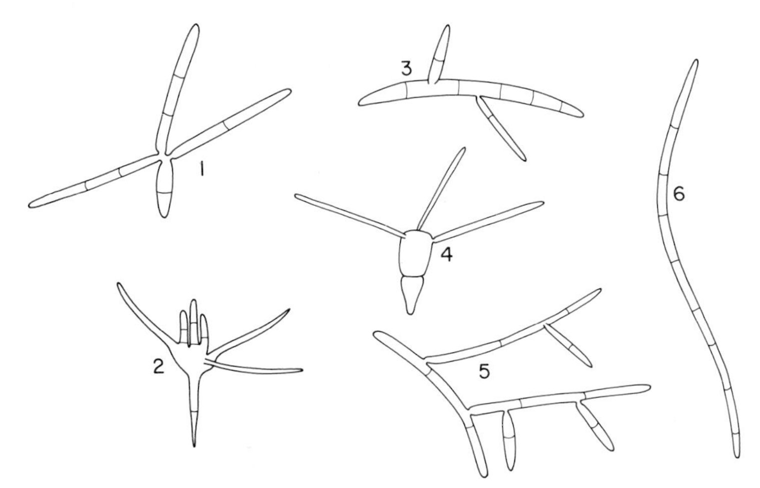

Figure 7.12. Spores of aquatic hyphomycetes illustrating the range of forms. 1. *Articulospora tetracladia*; 2. *Tetracladium setigerum*; 3. *Tricladium splendens*; 4. *Clavariopsis aquatica*; 5. *Varicosporium elodea*; 6. *Anguillospora longissima*. Not drawn to scale.

A very important epipelic community occurs at the boundary of fresh and salt water in estuaries and salt marshes. Filamentous algae and Cynanophyta are widespread and common (*Vaucheria, Enteromorpha, Ulothrix, Anabaena* and *Calothrix*) and are important in stabilizing mud. There are zones down the marsh, related to tide levels, both in macroscopic plants and in micro-organisms. Thus at the top *Navicula, Nitzschia* and *Rivularia* and occasionally *Nostoc* are common while *Caloneis, Pleurosigma* and different species of *Navicula* and *Nitzschia* occur lower down in the intertidal with *Microcoleus* and *Phormidium*. The Cyanophyta on mud flats make a significant contribution to the nitrogen balance in estuaries.

Bacillus, especially chromogenic species, are common on old, stable salt marshes and Gram positive bacteria in general are more common than Gram negative, a reversal of the normal distribution in water. Young, developing marshes have more Gram negatives. There is no detailed information on the distribution of other genera in relation to position on the marsh, but numbers are generally greatest near the top (Fig. 7.13).

The dominant species of fungi are probably more akin to a terrestrial than a marine flora (e.g. *Aspergillus, Penicillium*), though specifically marine yeasts do occur (*Trichosporon* and *Pichia* spp.). Fungal numbers also decrease down the marsh (Fig. 7.13). Both fungi and bacteria are important in the decay of marine detritus in the littoral zone (see page 112) and there are specialized communities based on the detritus. One of the best proved cases of natural heterotrophic growth of algae is also found here; *Nitzschia putrida* is a colourless diatom living in rotting seaweed. Ascomycetes (e.g. *Lulworthia*) are common on wood in the detritus and on structural timber exposed to water and they cause economically important soft rots. Detritus and particulate organic matter are important sources of carbon for ciliates and filter feeders in estuaries where the primary production in the plankton may be low because of turbidity.

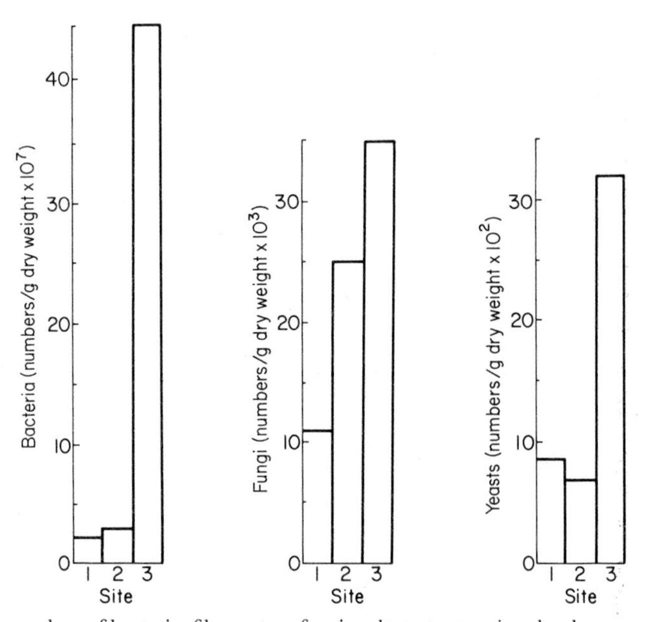

Figure 7.13. The numbers of bacteria, filamentous fungi and yeasts at various levels on a salt marsh, derived from dilution plate data. Site 1 was mud at low water mark, 2 was half way up the marsh and 3 was a *Spartina* bank at the top of high water. (Unpublished. Gillian Herrod-Hempsall and R. Campbell, Department of Botany, Bristol University.)

Complex communities of heterotrophs, phototrophs and chemotrophs may develop in some muds and these have been particularly studied in estuaries, though they occur elsewhere. Muds rich in organic matter, where only the surface is aerobic have colonies of algae, (e.g. *Navicula*) and Cyanophyta (e.g. *Lyngbya*) on the surface. Below these are the purple and green sulphur bacteria, oxidizing reduced sulphur compounds produced by *Desulfovibrio* in the deeper anaerobic parts of the mud (see Fig. 5.4 for the differing requirements of these organisms). The algae, blue-green algae and photosynthetic sulphur bacteria all require light but their absorption spectra are different (Fig. 7.14), the bacteria using light on either side of the algal maximum at 600 to 700 nm. They complement each other in the utilization of available light rather than compete for it. This also applies to the sulphur bacteria living in the hypolimnion below the algae in lakes (Fig. 5.3).

Pennate diatoms, and probably dinoflagellates, euglenoids and Cyanophyta, move up and down through the surface layers of the intertidal mud. As with those on sandy shores they usually appear as the tide recedes in daylight and retreat into the mud as the tide returns. The primary stimulus seems to be tides rather than light and indeed the upward migration of some (e.g. *Surirella*) is inhibited by light.

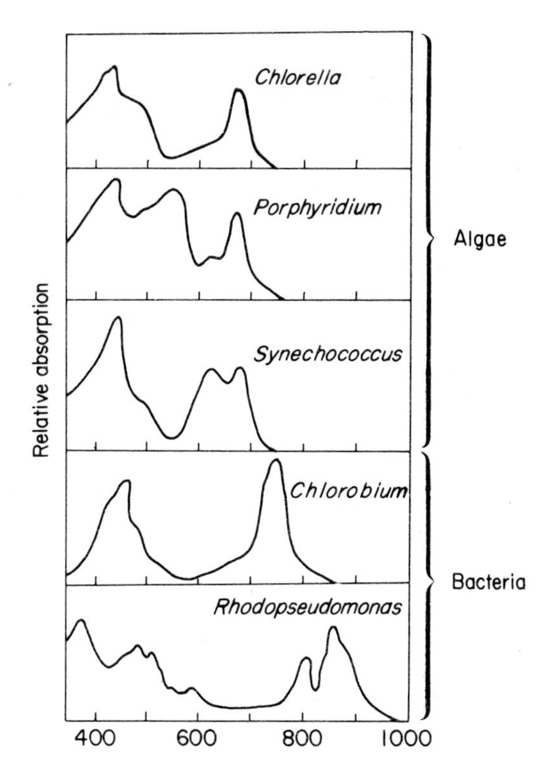

Figure 7.14. The absorption spectra of algae and Cynanophyta, with complementary use of the light by the sulphur bacteria growing below them in muds and water. Wavelength given in nm. (From R.Y. Stanier and G. Cohen-Bazire (1957). The role of light in the microbial world: some facts and speculations. In *Microbial Ecology*. 7th Symposium of the Society for General Microbiology. Cambridge University Press. p. 56–89.)

In the subtidal region algae occur on the sediments right down to the limits of the photic zone and diatoms seem to be the most important group in temperate waters, with counts of 50 to 100 cells/mm^2, though in the tropics blue-green algae such as *Lyngbya* and *Schizothrix* are more common. Many algae are found in deeper sediments but most of these have probably settled from the plankton. Fungi such as the thraustochytrids occur but are not thought to be important. The main micro-organisms in the deep ocean are bacteria, and they are predominantly Gram negative forms such as *Pseudomonas* and *Flavobacterium* with *Clostridium* under anaerobic conditions. Their numbers are very variable, saprophytic viable counts are usually 10^4 to 10^7 per g wet weight, with lowest values in the deepest water and, as in fresh water, the numbers decrease rapidly with depth into the sediment. Bacteria are especially important in recycling nutrients in anaerobic sediments, though even here protozoa may be nearly equal to them in biomass, if not in activity.

Epiphyton

The plant surface community varies from photolithotrophs which affect the host plant only by shading, through photo-organotrophs who supplement their photosynthetic activity with exudates, to heterotrophs and parasites which are entirely dependent on the exudates or the living host. Almost all aquatic plants have an epiphyton which may be very dense (Fig. 7.15A), though very rapidly growing plants remain 'clean' for a time and some species remain uncolonized

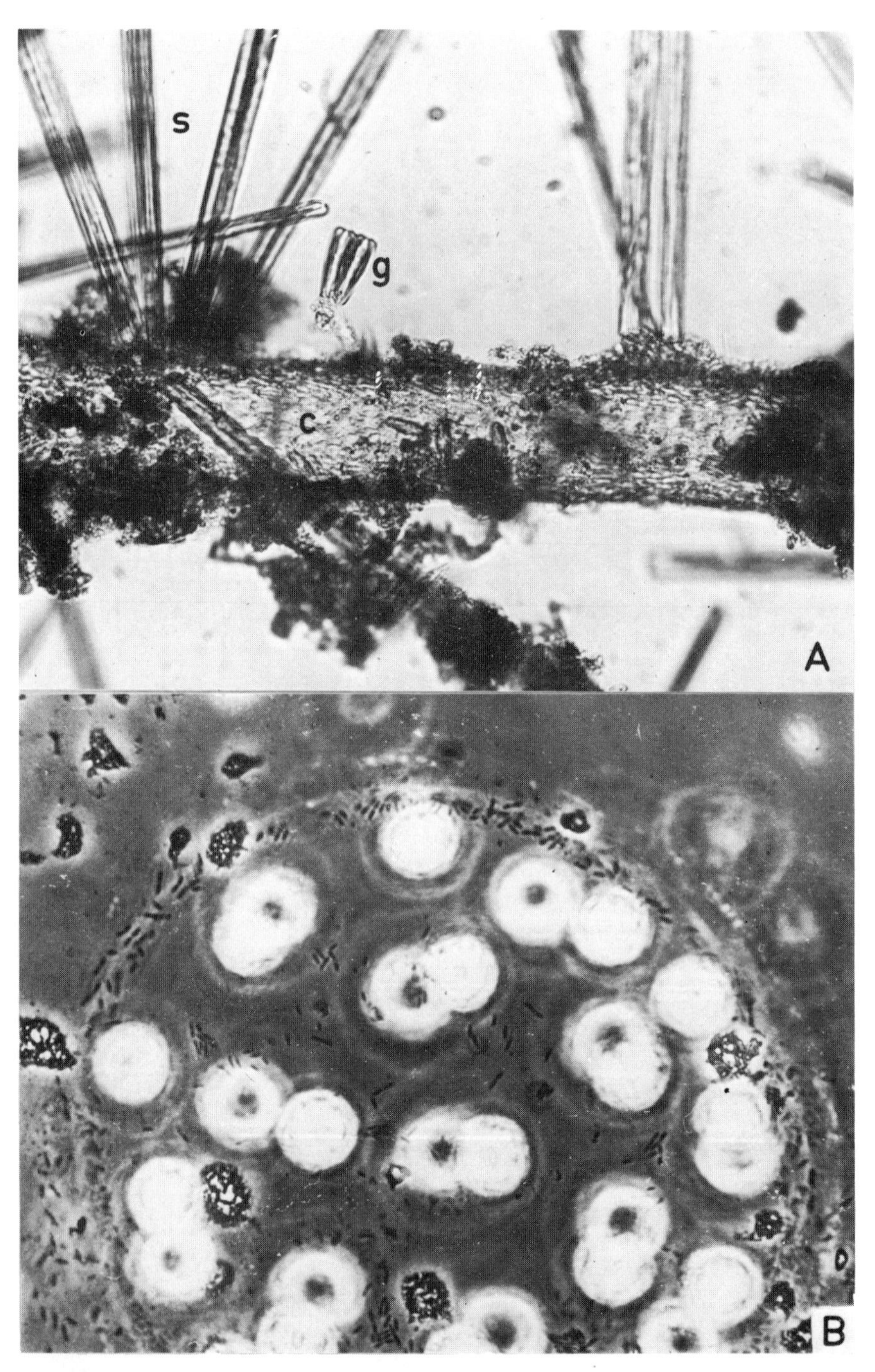

Figure 7.15. Epiphyton. A. Dense growth, mostly of *Synedra* (s), on the filamentous alga *Cladophora* (c). *Gomphonema* (g) and many smaller forms of diatoms and bacteria also occur in the mucilage around the filament. × 230. B. Bacteria in the mucilage of the planktonic green alga *Eudorina*. × 730.

permanently, probably because of the production of antibiotics (e.g. *Cyclotella* and *Stephanodiscus* are two diatoms which do not have a bacterial epiphyton). The production of such substances may also account for the specificity of some host/epiphyte associations but surface characteristics of the hosts, such as the amount of mucilage, are also important.

Planktonic algae usually have bacteria attached to or occurring within their slime sheaths (Fig. 7.15B) and sometimes fungi and protozoa are also present. A large part of the bacteria in the plankton may be epiphytic rather than truly planktonic except in very nutrient-rich waters. The role of the bacteria is uncertain, they are most probably saprophytes causing no harm, though *Cytophaga* can lyse algae. Many of the fungi are definitely parasitic, particularly chytrids (page 103, Fig. 7.6B and further reading 4). *Rhizophidium planktonicum* on the diatom *Asterionella formosa* may delay the time of maximum numbers and reduce the size of the bloom. At low populations of the host the rate of parasitism is low. Chytrids are most common in fresh water and thraustochytrids in the sea: both are cosmopolitan though little is known of the distribution of individual genera or species. Some of them have a limited host range, *Phlyctidium* grows on both *Scenedesmus* and *Pediastrum* but others are species specific, *Amphicypellus elegans* living only on dead or dying cells of *Ceratium hirundinella*.

Epiphytes on benthic algae, including macrophytes, and on benthic angiosperms, are very common. The dominant organisms (Fig. 7.3) are diatoms of two sorts, either adpressed forms attached by mucilage (*Cocconeis, Amphora, Epithemia*) or cells with a definite basal attachment pad and sometimes a stalk (*Achnanthes, Cymbella, Gomphonema, Synedra*). There are also filamentous epiphytes (*Oedogonium, Bulbochaete, Cladophora, Oscillatoria* and *Lyngbya*) which may themselves have epiphytes on them.

Bacteria are ubiquitous on benthic plants and they may live in mucilage, as in plankton, or they may be stalked forms (*Hyphomicrobium* and *Caulobacter*) or filamentous (*Leucothrix mucor* in the sea and *Sphaerotilus* in fresh waters). Very high numbers are recorded, especially on marine algae such as *Macrocystis* (10^4 to 10^9/cm^2 of *Vibrio, Flavobacterium* and *Pseudomonas*). On *Laminaria* numbers vary from 10 to 10^5 bacteria/cm^2, depending on the season of the year and the position on the frond in relation to its meristem and its disintegrating tip (further reading 1). Presumably these are mostly saprophytic, but some may be parasites and some are beneficial in that they probably supply the host with vitamins.

The large marine algae are host to a number of ascomycete fungi, some of which are parasites. There are also a number of strange organisms, with both fungal and protozoan affinities, the Labyrinthulales, which have motile cells moving within a network of slime threads. They are parasites of eel grass (*Zostera*) and probably also of algae. They are implicated, along with physical and climatic conditions, in the serious loss of plant cover which results in erosion of mud flats.

Associated with the epiphyton are free-living organisms within the protection of the epiphytic growth. They also occur in masses of floating filamentous algae, though they are not attached, and in mats of filamentous bacteria such as *Beggiatoa*. The flora consists of motile forms such as *Euglena, Navicula, Nitzschia* and many bacteria and protozoa. There are successions of protozoa in some of these communities, for example in *Beggiatoa* mats the initial colonization is by ciliates (*Colpidium*) eating the free-living bacteria, then predatory ciliates like *Linotus* reduce the *Colpidium* population and *Glaucoma* and *Chilodonella* then start to consume the *Beggiatoa* itself. The resulting detritus may then be colonized by *Paramecium* and *Coleps*. It is likely that successions of this sort are common in other natural communities for they are invariably seen in crude cultures of protozoa made in the laboratory from natural water samples inoculated into hay infusion etc. As in so many areas of protozoan ecology, more information is needed.

Epizoic communities

Most of the epizoic communities are similar to those on plants, merely using animals as a surface to grow on. Thus on whales, turtles, mollusc shells etc. there may be a dense covering of algae, bacteria and protozoa very like the fouling growths on ships. The copepods, arthropods and crustacean larvae in the zooplankton also have associated bacteria, chytrids and algae. Fish have a dense bacterial skin flora (1000 to 10,000/cm^2) and many fish also carry protozoa (e.g. *Trichodina*) on their skin and gill plates. Some deep sea fish have symbiotic bacteria (*Photobacterium*) which are luminous and provide light spots for lures and recognition patterns in those regions of perpetual darkness.

There are many parasitic bacteria and fungi which cause diseases of aquatic animals, perhaps the best known of which are the fungal diseases of fish caused by *Saprolegnia* and *Achyla* in fresh water. Sporadic outbreaks of economically important diseases of mussels and oysters also occur.

FOOD CHAINS

In most of these interrelated communities micro-organisms are the basis of all the food chains to higher animals. Even where the larger plants are the main primary producers, in littoral regions, the food chains are based on the microbial breakdown of the detritus that is produced, rather than grazing on the plants direct. This is particularly obvious in the seaweed beds, where the only direct grazing is by sea urchins, but which are very important and productive habitats (Table 3.1). Some of the interrelations between the different communities and the part they play in food chains is summarized in Figure 7.16.

EUTROPHICATION AND POLLUTION

The nutrient levels in natural waters are largely determined by the nutrient status of the surrounding land or of the rivers flowing into lakes or oceans. Normally the nutrient concentrations are not subject to rapid, permanent changes. Oligotrophic lakes, with productivities of 7 to 25 g C/m^2/year, normally remain in an oligotrophic state over historical time periods through there may be slow changes over a geological time scale. Seral changes, from lakes to marshes to dry land, result from an accumulation of sediment and are often not accompanied by a change in the nutritional status. Similarly eutrophic lakes, with productivities of 75 to 250 g C/m^2/year maintain a stable nutrient balance under most condi-

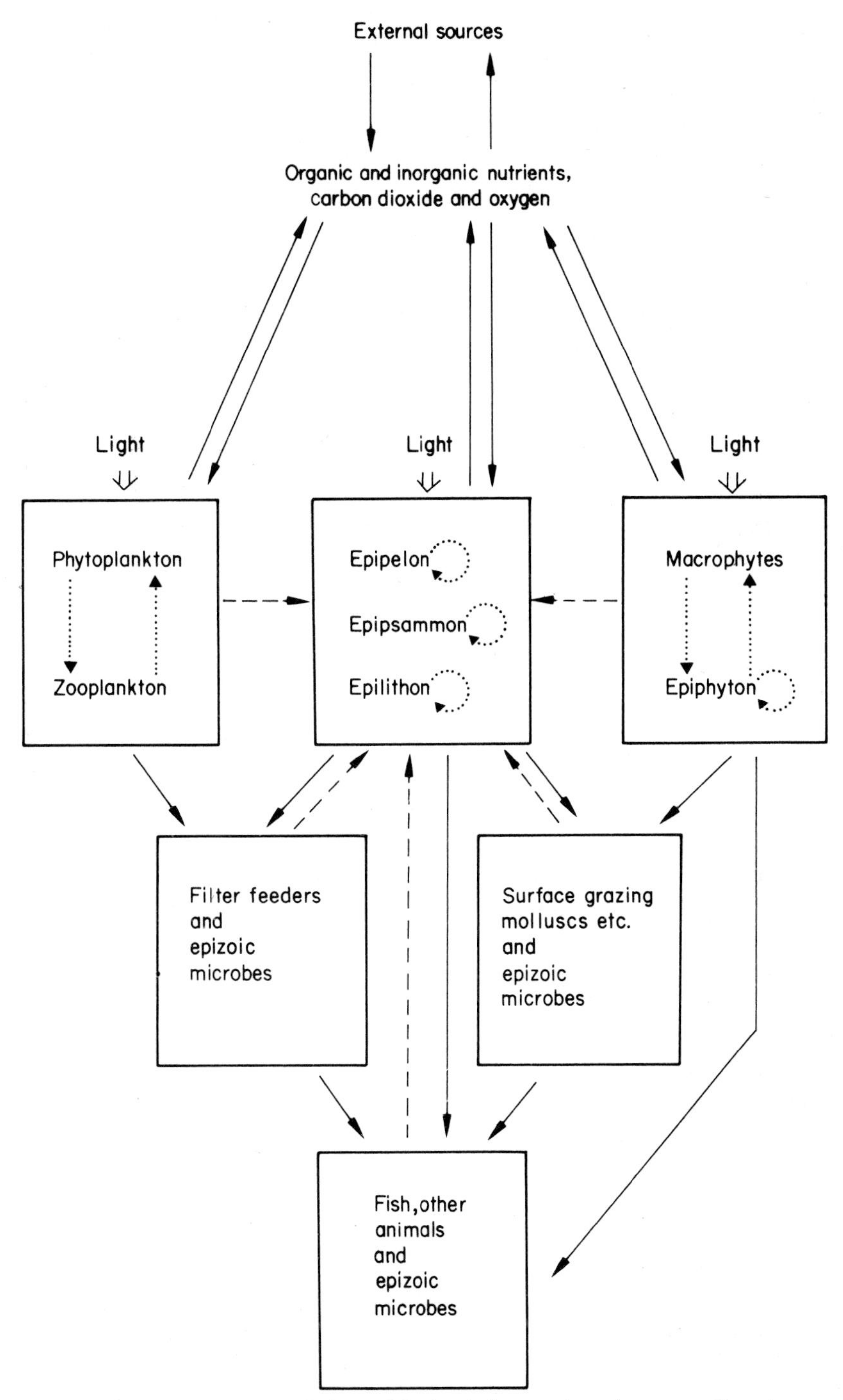

Figure 7.16. Community relationships and food webs in aquatic environments. Dotted arrows are internal nutrient cycles and food chains within a community, dashed arrows indicate the transfer of detritus between communities and solid arrows the transfer of other materials and energy between communities.

tions. Eutrophication is the comparatively rapid increase in the supply of nutrients, especially nitrogen and phosphorus, which are then maintained at the new high levels and result in an increased growth of algae and therefore of other forms dependent upon them (further reading 6). Eutrophication may be brought about by natural causes such as gross changes in the vegetation of a watershed caused by fire, or contamination by excreta of birds and animals near seabird rookeries or water-holes. In recent years the most common cause of eutrophication has been the addition of sewage (even if treated to remove organic matter), detergents rich in phosphate and agricultural fertilizers in run-off from fields. There is nothing intrinsically wrong with the eutrophic condition, the lake may be different from its former state but is not necessarily 'worse', unless the change is so abrupt, or the quantities of material involved so large, that the nutrient cycles break down or some economically important products, such as fish, disappear.

The effect of eutrophication on the microbial populations depends on the particular water concerned, the chemical nature of the additions and on what limiting factors are operating. If, for example, nitrogen and phosphorus are added when they are limiting, there will be an increase in numbers of bacteria, algae and protozoa: however their addition to a silicon limited diatom bloom will have little immediate effect. The composition of the population will change as the water becomes more eutrophic (page 99) and phosphate in particular often stimulates green algae and Cyanophyta, though other groups or species may be reduced (e.g. *Asterionella* was reduced in Lake Windermere as the phosphorus level increased). At high nutrient levels repeated blooms of different organisms may occur. These remove, at least temporarily, the available nutrients from the water, but they are released again during decay and support the next bloom. The micro-organisms themselves or their decomposition may produce taints and odours in water used for public supplies, and the high populations which occur under eutrophic conditions also cause problems in the premature blockage of filters in water treatment works.

At extreme nutrient levels the population is so dense as to be limited by mutual shading, and though the biomass may be great, the productivity per organism may be very low. Large amounts of decomposing material result in anaerobic conditions, slowing the decomposition rate. When the blooms reach this level then they are equivalent to pollution, damaging the environment by the addition of excess organic matter.

Pollution by organic matter has been extensively studied, particularly in rivers, in relation to the addition of sewage with a high biological oxygen demand. Small amounts of organic matter increase the numbers of bacteria, particularly those which are free suspended rather than attached to particulate material. There is also an increase in the epilithic algae, especially filamentous forms such as *Spirogyra, Ulothrix* and the blue-greens *Lyngbya* and *Phormidium*; each of these may have a dense epiphyton of diatoms (*Nitzschia palea, Diatoma vulgare, Fragilaria capucina* and *Meridion* sp.). The productivity of such communities may be very high (e.g. 350–700 g C/m^2/year). As the amount of organic matter increases the 'sewage fungus' may also occur: actually this is an assemblage of organisms including the fungus *Leptomitus lacteus* and the bacterium *Sphaerotilus natans* with associated protozoa, especially ciliates such as *Colpidium*. At high levels of organic matter pollution the water becomes anaerobic and there is a dense growth of *Sphaerotilus*, other anaerobic bacteria and ciliates. Provided that the only pollutant is organic matter then the water will recover its normal populations as the carbon is removed, as carbon dioxide, and the water is reoxygenated (Fig. 7.17; further reading 6). This form of pollution is therefore a problem only on a relatively local scale, unless the quantities added are extremely large. Furthermore rivers naturally receive large, and fluctuating, inputs of organic matter, from such events as leaf fall in temperate regions, which cause large oxygen demands. Though the organic matter is removed from the river by these recovery processes minerals remain and may cause eutrophication. One further aspect of pollution by sewage is the introduction of human pathogens into the environment, including bacteria causing food poisoning, typhoid, dysentery etc. (*Salmonella* spp., *Shigella* spp.) and also some *Clostridium* spp. which produce toxins. These microbes should not survive any reasonable sewage treatment scheme, but when raw sewage is discharged into a river or the sea they may persist for some time, despite the fact that saline water is mildly biocidal to the usual test organism *Escherichia coli*. The pathogens may be taken up by filter-feeding shellfish, and they may remain viable within these organisms.

The effects of chemical pollutants have already been considered (Chapter 3, pages 29–32). The variation in toxicity of a single compound to different organisms can have great effects on the community structure. This is particularly marked in the case of oil spills and the subsequent use of dispersants which are more toxic to the consumers than to the algal primary producers. This results in

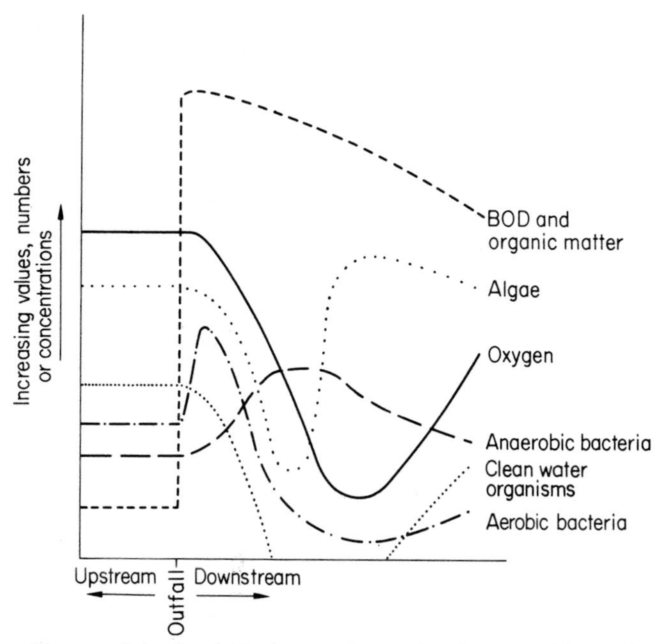

Figure 7.17. Diagram of the changes in the organisms and environmental factors in response to overloading with organic material from a sewer outfall into a river. The precise levels of the various factors and the distance the effect extends down-stream will depend on the quality and quantity of the effluent and on the flow rate of the river.

a removal of the grazing pressure on marine epilithic algae, and rocks which are normally almost bare become covered with a dense algal growth. This illustrates the importance of grazing in the control of normal population levels.

Heavy metals may be released into the aquatic environment intentionally, as in the use of copper to kill algal blooms in reservoirs or the application of biocides to prevent decay or the growth of fouling organisms. Some heavy metals may be metabolized, the most dramatic case being the production of methyl mercury by bacteria which makes the mercury more soluble and it then enters food chains.

Thermal pollution occurs in inland waters and on coasts, usually caused by the outlets from power station condensers. The populations are similar to those of natural hot springs (page 115) most noticeably in the increase in the importance of Cyanophyta and a reduction in species diversity, especially on the nanoplankton.

Inland waters have been polluted to an all too obvious extent ever since the Western countries centralized their industrial capacity in factories and towns. The sea has been regarded as an almost limitless dump. However the concentration of toxins, even from very dilute solution, by micro-organisms at the start of food chains means that this is not a safe assumption. Furthermore the most heavily polluted regions are usually the shallow coastal waters and salt marshes which have a high productivity compared with the open ocean, and so have an importance greater than their area would suggest.

FURTHER READING

1 ANDERSON J.M. & MACFADYEN A. (Eds.) (1976) *The Role of Terrestrial and Aquatic Organisms in Decomposition Processes.* 17th Symposium of the British Ecological Society. Blackwell Scientific Publications, Oxford. pp. 474.
A collection of reviews.

2 COLWELL R.R. & MORITA R.Y. (Eds.) (1974) *Effect of the Ocean Environment on Microbial Activities.* University Park Press, Baltimore, London and Tokyo. pp. 587.
Reviews and original papers.

3 HUTCHINSON G.E. (1957) *A Treatise on Limnology.* Vols. 1, 2 and 3. pp. 1015, 1115 and 660. Wiley, New York and London.
A comprehensive text on the chemical and physical factors and the populations, particularly algal plankton, in lakes.

4 JONES E.B.G. (Ed.) (1976) *Recent Advances in Aquatic Mycology.* Elek Science Publishers, London. pp. 749.
The only advanced text on fungi in water.

5 KINNE E. (Ed.) (1970) *Marine Ecology.* Vol. 1. Wiley Interscience, London, New York, Sydney and Toronto. pp. 1774.
The effects of environmental factors.

6 MITCHELL R. (Ed.) (1972) *Water Pollution Microbiology.* Wiley, New York. pp. 426.
Review of many aspects of pollution.

7 NOLAND L.E. & GOJDICS M. (1967) Ecology of freeliving protozoa. In T-T. Chen. (Ed.) *Research in Protozoology.* Vol. 2, 216–266. Pergamon Press, Oxford.
One of the very few reviews on protozoan ecology.

8 RHEINHEIMER G. (1971) *Aquatic Microbiology.* Wiley, London and New York. pp. 184.
Mainly on bacteria.

9 RILEY J.P. & SKIRROW G. (1975) *Chemical Oceanography.* Vol. 2. 2nd Edition. Academic Press, New York and London. pp. 647.
Good for organic and inorganic nutrients in the sea.
10 ROUND F.E. (1977) *Algal Ecology.* (In press.)
Comprehensive and very detailed text.
11 WETZEL R.G. (1975) *Limnology.* Saunders, Philadelphia, London and Toronto. pp. 743.
Useful for nutrient cycles.
12 WHITTON B.A. (Ed.) (1975) *River Ecology.* Studies in Ecology Vol. 2. Blackwell Scientific Publications, Oxford, pp. 725.
Reviews, including micro-organisms.

8 The structure and dynamics of microbial populations in the air

INTRODUCTION

Of the environments that we are considering air is in many ways the simplest for it consists of a single phase, a gas, apart from condensed water vapour and dust. It is composed (by volume) of 78% nitrogen, 21% oxygen, 0.9% argon and 0.03% carbon dioxide with traces of many other gases, and very low concentrations of organic and inorganic nutrients. It only contains free water at irregular intervals.

Various layers are distinguished in the atmosphere up to a height of about 1000 km. We are mostly concerned with the layer nearest the earth, the troposphere, which extends to about 11 km in temperate regions and 16 km in the tropics. Apart from irregularities near the earth's surface there is a steady decrease in temperature with height (about 1°C per 150 m) until the top of the troposphere when the temperature starts to increase.

Surfaces exposed to sunlight have higher temperatures than the surrounding air and as they warm the air in contact with them they will set up convection currents which are of great importance in air movement. At night surfaces cool by radiation, particularly if there is no insulating cloud cover, and may be at a lower temperature than the surrounding air: cool air then collects near the ground with warmer air above and such a situation is called a temperature inversion and can restrict upward air movement.

The effect of wind is to produce turbulence around stationary objects, though almost all surfaces are surrounded by a layer of still air, the laminar boundary layer, which is caused by the friction between the air and the object.

Micro-organisms usually occur in air in small numbers compared with soil or water and they are very rarely metabolically active because of the low water and nutrient levels. There are fungal, bacterial and algal spores and protozoan cysts; vegetative cells of all groups do occur but rapidly die unless they are protected from desiccation. Particles such as spores sediment out of still air, at about 1 cm/sec for a 20 μm diameter particle, and they are therefore dependent on turbulence and convection currents to remain airborne. Air is mainly important as a transport and dispersal medium for micro-organisms.

AIR OUTSIDE BUILDINGS

Release into the air

Most micro-organisms have no special mechanisms enabling them to become airborne (though see further reading 3 and 5 for fungi), soil microbes may be blown on dust, sea spray may contain organisms from the plankton and neuston and many of man's activities accentuate these passive methods of becoming airborne by creating turbulence and disturbing soil and vegetation. Rain creates aerosols on impact and the droplets can contain microbes from the surface or from the air through which the rain has passed.

Microbial numbers and distribution

The microbial component of the outside air spora is dominated by fungi except in close proximity to concentrations of animals and buildings where bacteria may be more important. Numbers are of course very variable, but there are

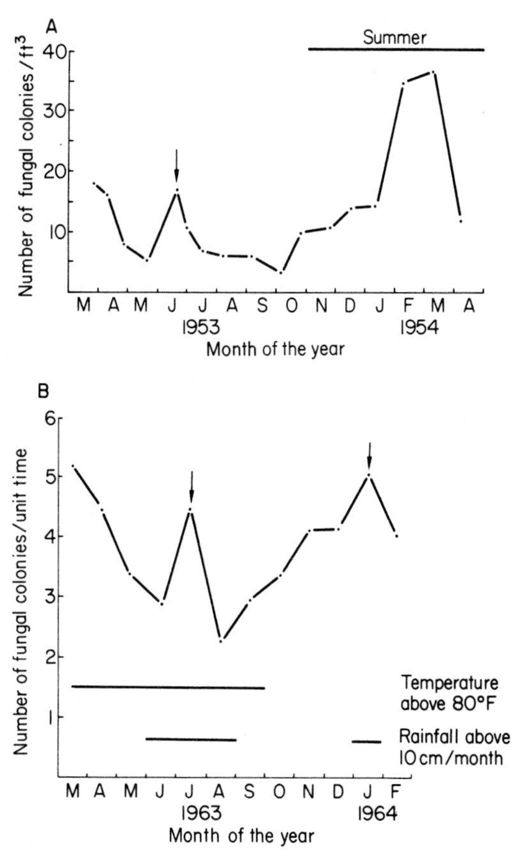

Figure 8.1. A. Annual variation in the fungal air spora over Dunedin, New Zealand; spores are most common in the warmer summer months. The high value indicated by the arrow was caused by a very high count (a spore shower) of one genus, *Penicillium*. (From M.E. di Menna (1955). A quantitative study of air-borne fungus spores in Dunedin, New Zealand. *Transactions of the British Mycological Society* **38**, 119–129.) B. Annual variation in the air spora of Hong Kong. Periods of high temperature (above 80°F) and high rainfall (above 10 cm/month) are shown: the very high temperatures reduce the air spora and wash out by rain may occur at the same time. The high values indicated by the arrows were caused by spore showers of *Aureobasidium*. (From P.D. Turner (1966). The fungal air spora of Hong Kong as determined by the agar plate method. *Transactions of the British Mycological Society* **49**, 255–267.)

several hundred bacteria per m^3 in the country and sometimes several thousand/m^3 in cities. Fungal spores are more common, and tens of thousands/m^3 are possible under summer conditions over agricultural land. These figures may rise much higher near large local spore sources and conversely drop over the oceans by 2 or 3 orders of magnitude. The two commonest fungi are the deuteromycete *Cladosporium* and the basidiomycete yeast *Sporobolomyces*. Basidiospores are also very common.

The density of the spora varies on an annual and a daily basis. The annual variation is mainly caused by the climate for in temperate countries the lower winter temperatures are less favourable and there is also increased wash out by rain (Fig. 8.1A). Rain and snow are quite efficient at impacting and removing spores greater than 2 μm diameter from the air. In tropical conditions, the temperature may be so high as to cause desiccation and the levels drops during both prolonged dry conditions and monsoon periods (Fig. 8.1B). Arctic and Antarctic air is generally low in spores because of the lack of local production.

Diurnal fluctuations are mostly related to the means of spore discharge and various patterns have been distinguished (Fig. 8.2; further reading 2). Basidiospores, particularly colourless ones (including *Sporobolomyces*), are mostly present at night. *Phytophthora* sporangia are released as the relative humidity drops after dawn and they reach their peak numbers in the air during the morning. *Cladosporium* and *Alternaria* depend on turbulent air and are most common around noon. The formation of *Erysiphe* spores is apparently controlled by light and they only reach maturity at midday and are then removed by turbulent air in the afternoon. These examples are all 'fine weather spores' which are not dependent on rain for release. The 'wet air spora' contains many ascomycetes, especially those with perithecia which require a thorough wetting before discharge, though heavy dew near dawn may be sufficient, so they may also be common in the very early part of the day. The effect of rain is therefore to first diminish the spora by wash

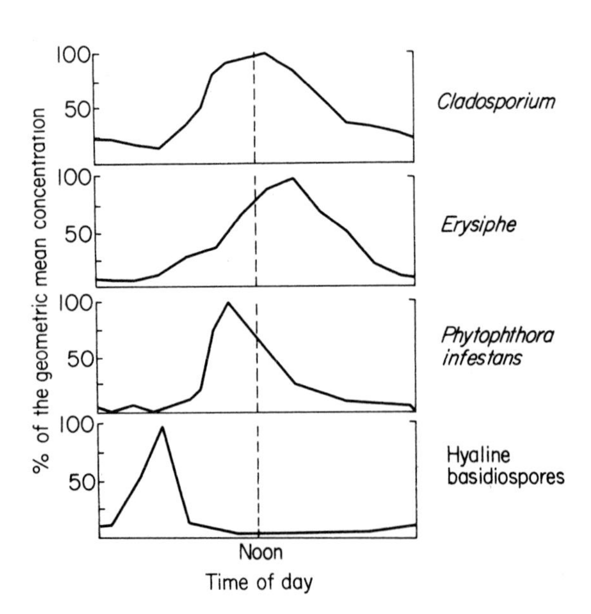

Figure 8.2. Diurnal variation in spore concentration for different genera or spore types, showing different patterns of maxima. The concentrations are expressed as a % of the geometric mean concentration for several samples. Hyaline basidiospores are mainly *Sporobolomyces*. (From J.M. Hirst (1953). Changes in the atmospheric spore content: diurnal periodicity and the effect of weather. *Transactions of the British Mycological Society* **36**, 375–393.)

out and then to increase it as the wet air spores and splash dispersed species become temporarily important.

The diurnal rhythm can be distorted if the main source of spores is some distance from the sampling point so that the travel time puts the spore peaks later than expected. This may occur with 'country spores' sampled in the middle of large cities or even more noticeably over large bodies of water where the spores may be days away from their source. The sea itself contributes algae and a relatively low number of bacteria (1 to 5/m^3) but almost all the fungal spores come from land. Spores may be randomly dispersed over a wide area if turbulent conditions prevail, but if the air flow is relatively laminar spores may travel as clouds for a considerable distance, each cloud representing a particular source of inoculum or release at a particular time of the day (Fig. 8.3). A few fungal spores can be found even in the middle of the oceans.

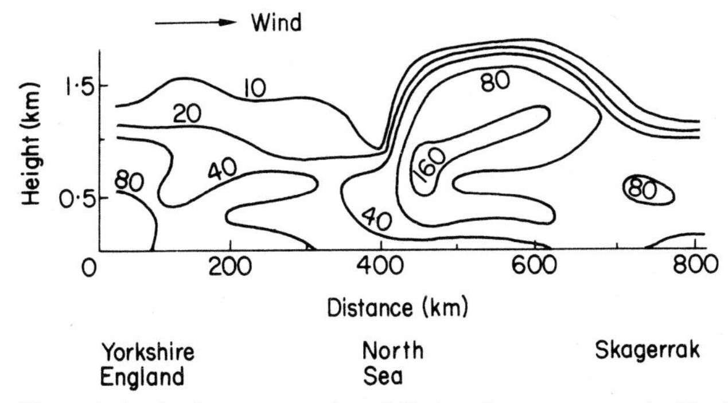

Figure 8.3. The variation in the concentration of *Cladosporium* spores over the North Sea with the height and distance from the coast of England. The spore concentration is in hundreds/m^3 and points of equal concentration are joined by 'isospore' lines. (From J.M. Hirst, O.J. Stedman and G.W. Hurst (1967). Long distance spore transport: vertical sections of spore clouds over the sea. *Journal of General Microbiology* **48**, 357–377.)

The long distance movement of spores is of great importance for the spread of plant pathogens though generally the oceans are an effective barrier because the inoculum level would be so low after passage that infection would not occur. It is, however, quite possible for spores to cross short stretches of water like the English Channel or the North Sea. The distance travelled depends greatly on the air layer in the atmosphere which is involved. If the spores only get into turbulence near the ground they may be carried only a few metres: 90% of spores are deposited within 100 m of the source. If the spores are carried up by convection currents or into weather conditions involving rising air masses on frontal systems, they can be dispersed over tens or even hundreds of kilometres. A good example of this is the spread of rust spores in the U.S.A. The rust, *Puccinia graminis*, does not overwinter in the very cold prairie states but winter wheat near the Gulf of Mexico remains infected. Direct observation shows that uredospores, produced in early summer, travel up the Mississippi Valley on the southerly winds for over a thousand kilometres. The source of spores can be recognized by identifying their physiological race. Spore clouds have also been followed from known outbreak centres to known new infections by inference from information on the weather systems over the flight path at the time. It is now clear that long distance dispersal in a single stage is a regular feature with some pathogens. Plant pathogens may also travel in a series of shorter jumps, landing and going through a spore cycle on crops between the outbreak centre and the final destination.

Care is required in translating spore trap data to infection by pathogens: some of the spores may not be viable e.g. many hyaline basidiospores do not travel far in a viable state. Even if present and viable, the numbers may be insufficient to cause infection. Hosts must be susceptible to the pathogens and leaves are not usually very efficient at trapping spores: the tips turn downwind and present the leaf edges to the airstream. Leaves are not sticky, as are artificial collecting surfaces, and spores may bounce off even if they penetrate the laminar boundary layer. Hairy leaves have a deep boundary layer and so are inefficient traps, but leaves with a few very long hairs may be quite efficient for the hairs have a very small diameter and may stick up above the boundary layer.

If plant pathogens can travel long distances there is no reason why the same air masses should not carry other micro-organisms, though detailed data are not common. Foot and mouth disease virus may travel for 2.5 to 3.0 km and there are reasonable data for up to 120 km, with enough viable particles left for infection under moist, cool conditions (further reading 4). However the transmission of animal diseases is not usually important in outside air.

Apart from long distance dispersal, by air or other means, plant pathogens usually have short range spread (further reading 2) within the infected crop which may be by a different mechanism or by a different spore stage. It is important to know these methods to formulate suitable control procedures, predict outbreaks of disease and apply protectant fungicides before the disease is evident on the crop. Thus *Phytophthora infestans* may arrive in a field by planting infected tubers or by air borne sporangia but thereafter it may spread from these infection foci by means of zoospores, the production of which can be predicted by measurements of rainfall and humidity (further reading 3). Theoretical calculations and modelling approaches have been used to predict dispersal and calculations can give the maximum disease intensity, for example, if all the factors are working in favour of pathogen spread in that direction. Similarly a safe quarantine distance can be given in terms of 'at X km from source there is a specified chance of the disease occurring'.

The vertical distribution of the air spora has been mentioned in passing, in that the higher the particle the greater chance of long range dispersal. The density of spores usually decreases approximately logarithmically with altitude above the point on the ground where they originate and, even though turbulence tends to disperse spores as they rise, the spore clouds may be identified for some distance as we have seen. If spores are moving into the area from an outside source, numbers may increase with height (Fig. 8.3). The vertical distribution of spores will also reflect the local air conditions, particularly the amount of turbulence, the presence of temperature inversions and the strength of convection currents. Fair weather cumulus clouds which form by convection may have higher numbers of microbes than the surrounding air, reflecting the origin of the convection current from the local vegetation. Global weather patterns may also influence the vertical distribution of the air spora.

The upper air flora has recently been investigated in connection with space programmes, particularly the need to retain sterility of interplanetary probes. Dark spored fungi such as *Cladosporium* and *Alternaria* and bacteria like *Bacillus* and *Micrococcus* have been detected at 4 km in concentrations of 2 to $500/m^3$. By sampling very large volumes of air, spores have been detected as high as 30 km, but there are severe conditions of desiccation, ultra-violet radiation and low temperature at these altitudes in the stratosphere. High altitude winds may disperse such spores very long distances but whether they remain viable is doubtful.

Viable spores are mostly confined to the lower 6 km of the troposphere but even here there is a survival problem. The main factors are again desiccation and U.V., though there is sometimes the possibility of photorepair of radiation damage. Some bacteria also loose viability because of oxygen toxicity and an unknown mechanism, the open air factor, causes loss of viability at night: the open air factor is possibly a chemical air pollutant.

Microbes on aerial plant surfaces

Plants projecting into the air are comparatively nutrient-rich surfaces, with a slightly ameliorated microclimate, and micro-organisms growing on them are an important source of outside air flora. The other sources (soil and water surfaces) have already been considered (Chapters 6 and 7). Stems and twigs have a resident flora but this has not been extensively investigated: it includes many algae and Cyanophyta (e.g. *Chlamydomonas*, *Chlorella*, *Stichococcus*, *Tetracystis*, *Oscillatoria*) and lichens if the structures are perennial. Small twigs are quite efficient at trapping spores though in the absence of pathological slime fluxes or large numbers of algal primary producers the nutrient level is quite low. The flora of the leaf (the phylloplane), forms a distinct community (further reading 1, 6 and 7).

The humidity near the leaf surface is low and water may be limiting to growth, except under tropical conditions. Nutrients come from airborne dust and pollen and from leaf exudates. There is very little information on exudates. About 5 or 6% of the dry weight of leaves can be removed by artificial leaching of young leaves, though this can be regarded as a maximum figure for older leaves and species with thick cuticles would exude far less, especially under natural conditions where leaching is probably less rigorous. Leachates, dew, rain etc. from leaves contain minerals in small quantities but most of the nutrient is organic, particularly organic acids and simple sugars. Nutrients are often limiting for growth on leaves and the artificial application, of urea for example, results in an increase in the numbers of phylloplane micro-organisms.

The microtopography of leaf surfaces is very variable, depending on the plant species, and it has a great effect on the microbes because of the creation of microhabitats which may be more favourable than the general surface, e.g. shade and higher humidity under spines or leaf hairs (Fig. 8.4). The cuticle is usually coated with wax which varies in structure and chemistry with the species and the environmental conditions: often it forms a hydrophobic layer of tubules, plates or other shapes which makes it difficult for phylloplane organisms to

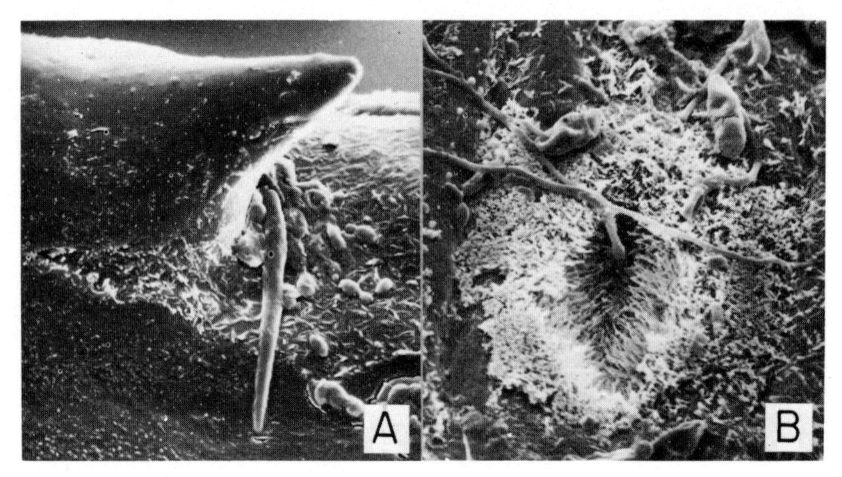

Figure 8.4. Stereoscan photographs of pine (*Pinus nigra* var. *maritima*) needle surfaces. A. Spine on the edge of the needle with yeast cells and a spore of *Lophodermella sulcigena* underneath it. × 1,200. (From R. Campbell (1972). Electron microscopy of the epidermis and cuticle of needles of *Pinus nigra* var. *maritima* in relation to infection by *Lophodermella sulcigena*. *Annals of Botany* **36**, 307–314.) B. Tubular wax in the epistomatal cavity with spores and hyphae of *Hendersonia acicola*. × 1,400.

actually contact the leaf surface. Wax also greatly affects the wettability of the leaf and is therefore important in controlling the amount of water on the leaf, the leaching from the leaf, splash dispersal of spores and the effects of fungicides, particularly when they are applied in aqueous carriers. The main constituents of wax are n-alkanes and their esters, which are resistant to decomposition by many of the phylloplane flora though some bacteria have the ability. The wax may specifically encourage the germination of some fungal spores, and so may be one of the many factors which affect the composition of the flora.

Cladosporium spp. (e.g. *C. herbarum*) and *Aureobasidium pullulans* seem to be ubiquitous phylloplane fungi. Yeasts are very common and the species isolated include *Sporobolomyces odorus*, *Cryptococcus* spp. and *Torulopsis* spp. Chromogenic and Gram negative bacteria, including species of *Xanthomonas*, *Erwinia*, *Flavobacterium* and *Pseudomonas*, are common and the Gram positive *Lactobacillus*, *Corynebacterium* and occasionally *Bacillus* occur. The types and numbers of phylloplane organisms are summarized in Table 8.1.

Pathogens, mostly fungi, need to grow on the leaf surface; even obligate parasites have to germinate and penetrate the leaf before they can obtain nutrients in any quantity. Reserves within the spore are important but the pathogens may also obtain soluble nutrients from the leaf, and pollen grains have been shown to increase infection by some parasites, probably by providing nutrients. Algae and blue-green algae occur on leaves under damp conditions: coccoid forms seem to be the most common. Moist tropical conditions also allow colonization by algae (e.g. *Trentepohlia*), lichens (often with *Trentepohlia* as the symbiont) and even protozoa have been recorded (further reading 7). The flora is probably more extensive in the tropics though information is sparse for most work has been on temperate plants. The number of organisms (Table 8.1) may

Table 8.1 The types and numbers of micro-organisms found on leaf surfaces of various plants. See also Table 8.2. (From various sources: 1. Jensen, V. (1971) The bacterial flora of beech leaves. In T.F. Preece & C.H. Dickinson. Eds. *Ecology of Leaf Surface Micro-organisms*. Academic Press, London & New York. 463–469. 2. Campbell, R. & Martin, M.H. Bristol University, unpublished data 3. Gingell, Sally M., Campbell, R. & Martin, M.H. (1976) The effect of zinc, lead and cadmium pollution on leaf surface microflora. *Environmental Pollution* **11**, 25–37. 4. Dickinson, C.H., Austin, B. & Goodfellow, M. (1975) Quantitative and qualitative studies on phylloplane bacteria from *Lolium perenne*. *Journal of General Microbiology* **91**, 157–166. 5. Ruinen, J. (1961) The phyllosphere. 1. An ecologically neglected milieu. *Plant and Soil* **15**, 81–109. 6. Bainbridge, A. & Dickinson, C.H. (1972) Effect of fungicides on the microflora of potato leaves. *Transactions of the British Mycological Society* **59**, 31–41.)

Micro-organism	Plant Species	Numbers micro-organisms per cm^2	Reference
BACTERIA			
Unidentified	Beech (*Fagus sylvatica*)	3,300–31,700†	1
	Strawberry (*Fragaria* sp.)	34,100	2
	Cabbage (*Brassica oleracea*)	5,200	3
	Rye grass (*Lolium perenne*)	1,000–250,000†	4
Actinomycetes	Rye grass	1–4	4
Azotobacter	*Cocao*	10×10^6*	5
Beijerinckia	*Cocao*	27×10^6*	5
FUNGI			
Unidentified filamentous	Strawberry	7,200	2
	Cabbage	450	3
	Potato (*Solanum tuberosum*)	13,200–38,500 μm/cm^2†	8
Unidentified spores	Potato	1,350–5,350†	6
Unidentified yeasts	Strawberry	5,200	2
	Cabbage	1,700	3
	Potato	1,340–224,600†	6
Cladosporium	Potato	760–2,900	6
Alternaria	Potato	50–150	6

† Numbers vary with time of year, normally increasing as leaf ages and becomes senescent.
* Tropical.

seem large but, as in the case of the rhizosphere (page 78), the area of leaf covered varies from 1% to perhaps 10% for temperate conditions, but higher on tropical evergreen plants.

Buds contain bacteria, yeasts and filamentous fungi and this site gives protection during overwintering. Their numbers are too small for their reproduction to keep pace with leaf expansion in spring, so newly opened leaves under natural conditions have a very sparse flora. They acquire most of their microbes from the atmosphere. Numbers increase during the season due to this inoculation and to the growth of organisms; maximum populations occur as leaf senescence starts or when the leaf is diseased. Such leaves may produce more exudates. Sometimes a succession can be demonstrated on leaves, which may be related to leaf age and to the availability of inoculum in the air (Fig. 8.5). Some fungi like *Aureobasidium* are present in buds, on the leaves at all ages and even inside the leaves, though they cause no pathological effects. *Cladosporium* colonizes the leaves after emergence and reaches its maximum in late summer. Other fungi occur sporadically or are present in small numbers. Though microbes may be isolated from leaves they may not be active; direct microscopic examination shows that *A. pullulans* is actively growing for only a few weeks at the beginning of the season and then is present as microsclerotia.

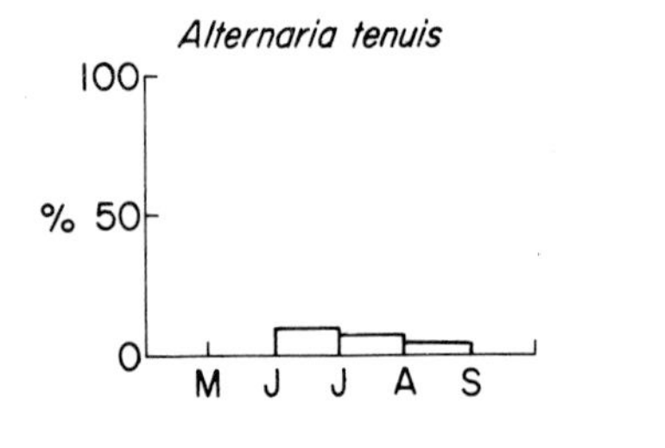

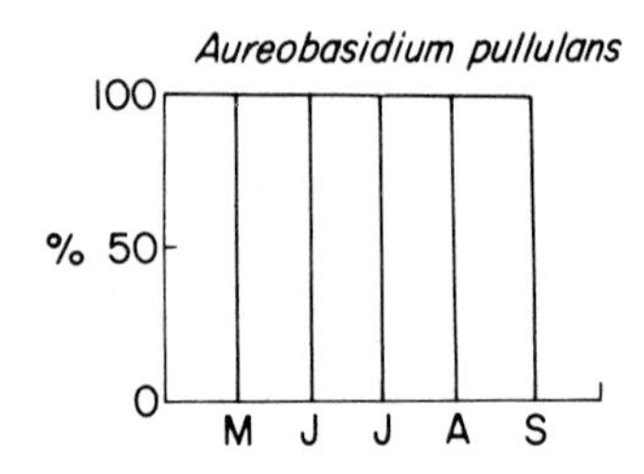

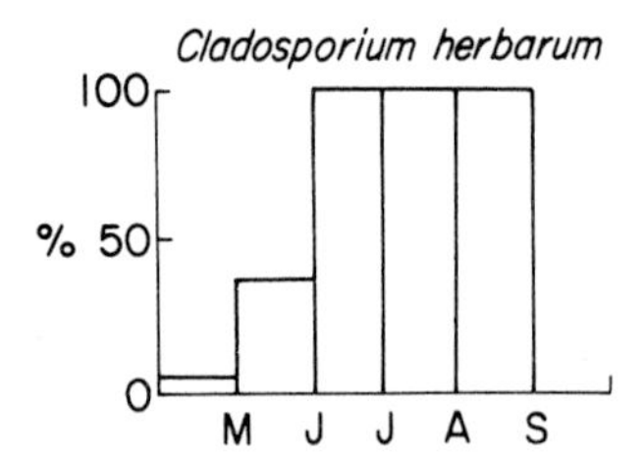

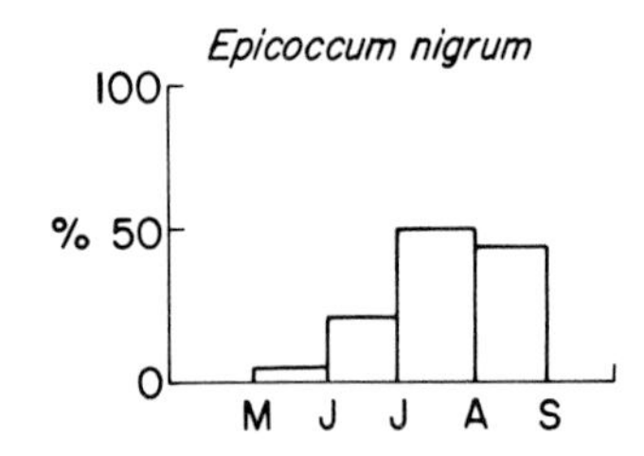

Figure 8.5. Percentage frequency of isolation of four phylloplane fungi on monthly samples of 50 washed leaf squares. (From G.J.F. Pugh and N.G. Buckley (1971). The leaf as a substrate for colonization by fungi. In: T.F. Preece and C.H. Dickinson. (Eds). *Ecology of Leaf Surface Micro-organisms*. Academic Press, London and New York. pp. 640.)

Competition occurs on the phylloplane for nutrients, and possibly also for favourable microhabitats, and attempts have been made to exploit this for the control of parasites by saprophytes. Rates of infection are greatest when pathogens are inoculated without saprophytes and infection can be reduced, in experimental conditions, by prior inoculation with saprophytes which colonize potential infection courts before the pathogen arrives. Saprophytes may also produce inhibitors to pathogen germination or growth, and stimulate the host plant defences (e.g. phytoalexin production). These effects have not yet been used on a commercial scale, but the phylloplane flora can be crudely manipulated by applications of simple nutrients, so the possibility exists of encouraging natural control by alteration of the environment with chemicals which are potentially less harmful than many fungicides. Fungicides themselves may produce unexpected results; for example with the *Alternaria* the systemic fungicide benomyl may cause increases in the disease on cauliflowers. *Alternaria* is known to be almost unaffected by the normal application rates of benomyl while the saprophytes would be reduced in numbers: it is tempting to link reduced saprophyte numbers with the observed increase in the disease. Wide spectrum fungicides and bacteriocides generally cause a decrease in the saprophytic flora but there may be resistant organisms. Many of the traditional fungicides, such as those based on copper, are phytotoxic at concentrations near to those required for control of the fungi and phytotoxic compounds can increase exudation by causing premature senescence. Conversely there is evidence that high microbial populations may increase the rate of senescence, and non-phytotoxic fungicides delay senescence, probably by their effects on the general microflora.

The effects of atmospheric pollutants have been studied, especially sulphur dioxide. In general it decreases the number of micro-organisms, including fungal pathogens. Thus, there may be increases in disease when clean air legislation reduces sulphur dioxide emission. Heavy metal pollutants, including dust from metal smelting industries on a local scale, and probably lead from car exhaust, also reduce total numbers (Table 8.2) and species diversity. As with most toxic agents there is often a selection of tolerant strains or species which then dominate the flora.

Table 8.2 The effect of exceptionally high levels of heavy metal contaminants near a smelter on the phylloplane micro-organisms of apple seedlings. 'Unknown yeast 1' and the *Aureobasidium* isolates were subsequently found to be tolerant to high levels of some heavy metals, and are presumably tolerant strains that became more important as the susceptible population declined. All differences between the control and contaminated site were significant at the 0.1% probability level. (Unpublished data, R. Campbell and M.H. Martin, Bristol University).

	Contaminated site	Control site
Heavy metal levels; ppm		
Lead	444.2	26.1
Zinc	658.1	37.2
Cadmium	11.5	2.0
Micro-organism numbers/cm^2		
Filamentous fungi	5,000	15,100
Bacteria	150	39,350
Yeasts*	7,200	16,900
'Unknown yeast 1'	6,300	2,450
Aureobasidium	4,850	2,300

*Excluding 'Unknown yeast 1' and *Aureobasidium*.

The phylloplane of Pteridophytes and Bryophytes has been much less studied than that of economically important crop plants, but they seem to have a similar fungal and bacterial flora to Angiosperms. Algae, particularly diatoms (e.g. *Pinnularia, Navicula* and *Hantzschia*) occur on moss fronds even under comparatively dry conditions. Under wet conditions there are more diatoms and also some Chlorophyta and Cyanophyta. The algae may be stratified with respect to the growing point (e.g. on moss in the Antarctic) and are most common in the region where the leaves of the moss begin to die. Coccoid Chlorophyceae and Xanthophyceae and the blue-green *Pseudanabaena* occur near the tips while diatoms and the blue-green *Nostoc* occur lower down (Fig. 8.6).

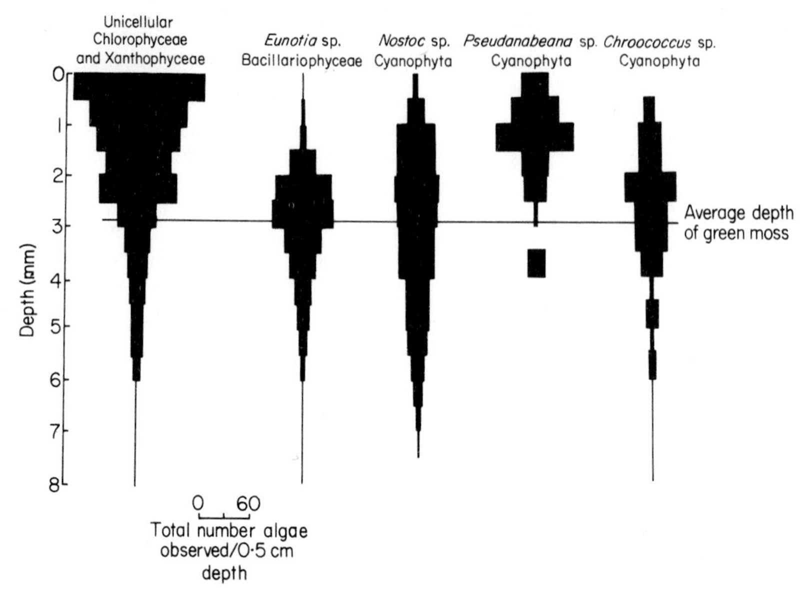

Figure 8.6. The relative numbers of algae and Cyanophyta on moss and liverwort plants on Signy Island, Antarctica. (Unpublished data, Dr Paul Broady, British Antarctic Survey.)

AIR INSIDE BUILDINGS

Microbial numbers and distribution

The main concern of studies of indoor air has been the spread of infection in hospital wards (further reading 4) though there is also general information on air movement (further reading 3). Inside air is very rarely still, even in unoccupied rooms there are draughts from windows and doors with velocities of 30 or 40 ft/min, upward convection currents from heaters and downdraughts by cold windows. People in the room cause convection currents, by the loss of about 105 kJ/h, and these can carry particles from the body surface of up to 50 μm in diameter. Moving people obviously create turbulence. One of the major causes of airborne dust in the house and hospital is bed making and house work, including 'dusting', also moves particles off surfaces and back into the air.

There are fungi in indoor air, though they are not so important as outside. The dominant genera are common saprophytes e.g. *Penicillium* and *Aspergillus* and others which may grow on food and damp wall surfaces. Fungi and actinomycetes are dominant in the air of barns containing mouldy hay, some forms of intensive animal rearing houses and mushroom farms. There can be extremely high concentrations and continued exposure may lead to allergic responses such as farmer's lung. However bacteria usually dominate the inside air though their numbers vary greatly depending on the situation. In well ventilated rooms there may be several hundred/m^3 but in slaughter houses and waste disposal plants the number may rise to tens or even hundreds of thousands/m^3.

In normal living conditions bacteria come from two main sources (further reading 3 and 4); firstly the surfaces of the upper respiratory tract from which they are released in aerosol particles in sneezing, coughing and talking. The liquid evaporates in a fraction of a second to leave droplet nuclei of about 10 to 15 μm, each of which carries only one or a very few viable organisms. Viruses are also released as aerosols in this way. The second major source of bacteria is skin squamules. Clothing reduces release into the air, though by no means eliminates it, and males often release more infective particles than females, especially from the perineal region. About 70 to 90% of household dust is skin squamules. The bacterial flora of indoor air reflects that of the skin and respiratory tract, being dominated by staphylococci (see below), though *Bacillus* spp. and endospores of *Clostridium perfringens* also occur frequently. In the presence of infected individuals there are other pathogenic bacteria including those causing tonsillitis, tuberculosis and whooping cough. Many viral diseases are also air transmitted, e.g. colds, influenza and German measles.

Airborne microbes indoors do not lose viability because of radiation (cf. outside air), but temperature is important and even 35 to 40°C can cause quite rapid inactivation. High relative humidities, though not saturation, prolong viability of aerosol particles and dust may have adsorbed water. Bacteria and viruses on dust can withstand normal indoor conditions for many weeks.

Microbiology of animal surfaces

Different species of animals have different flora which is mostly controlled by; 1. the body temperature; 2. the amount of fur, feathers etc. determining the surface temperature; and 3. the diet which influences the composition of the sebum and the sweat in those animals which have sweat glands. Sebum and sweat are the main nutrient sources for micro-organisms on human skin which has relatively few resident species, though it can pick up a lot of contamination from soil and from faeces. Detailed consideration of the human surface flora is outside the scope of this book (further reading 8), and will be briefly considered only as a source of indoor air spora. Fungi are represented mainly by the yeasts of which the best known is *Candida albicans*, a normal, apparently saprophytic, resident of many people which may occasionally become pathogenic when the competition from bacteria is relieved by the use of bacteriostatic agents e.g. wide spectrum antibiotics such as tetracyclines. Several other yeasts are very common, particularly *Pityrosporum ovale* on the scalp and *P. orbiculare* on glabrous skin. Some of the normally pathogenic dermatophytic fungi such as *Microsporum* and *Trichophyton* may also be present on the apparently healthy scalp. *Torulopsis galerata* and non-pathogenic *Candida* spp. occur widely, but not in large numbers, on the skin between the toes and in the axilla.

There are very few species of resident viruses, though many may be found as contaminants. *Herpesvirus hominis* occurs in a latent state and causes cold sores when conditions are favourable for infection.

The bacterial flora has been studied in relation to the infection of wounds, either natural or surgical, particularly by *Staphylococcus aureus*. Staphylococci are the dominant bacteria on the skin, though *S. aureus* is common only in the perineal region, on surfaces within the nose and on areas contaminated by nasal secretions. About 10% of the normal population carry *S. aureus*, apparently without harmful effects. *S. epidermidis* is ubiquitous and usually present in hundreds or thousands of cells/cm^2, though the density varies between individuals and with different positions on the body. It is most numerous near the mouths of hair follicles, and areas of moist skin e.g. the axilla, are most densely populated (10^4 to 10^5 cells/cm^2). *Sarcina* spp. are also common on healthy skin, especially in children. There is a substantial anaerobic flora in sebaceous gland ducts, *Propionibacterium acnes* is common and clostridia are also found. The scalp has a different flora, dominated by Gram positive rods (e.g. *Bacillus*) rather than the cocci of glabrous skin.

The reason for the comparatively restricted diversity of the flora on the normal skin is unknown. It is assumed that competition between residents is important, for when the balance is artificially disturbed by antibiotics or washing with soaps containing bacteriocides there are increases in normally unimportant species which are resistant to the agent employed. Some skin lipids are bacteriocidal and bacteriocins (antibiotics active against strains of the same species) may be important in maintaining the dominance of *S. epidermidis* over *S. aureus*.

In the mouth *Streptococcus salivarius* is most common but there is a specialized flora in the plaque on the surfaces of teeth which is dominated by Gram positive rods and filaments (mostly *Actinomyces*). The nasopharynx is dominated by Gram positive cocci and the occurrence of *S. aureus* has already been noted. The way in which this potential pathogen, and others in the nasopharyngeal region, is controlled so that apparently healthy carriers occur, is not known. Evidence from the administration of wide spectrum antibiotics again suggests that under normal conditions there is competition between the members of the resident flora so that pathogens, as alien species, do not become established.

The study of air flora has grown out of plant pathology and medicine where the pathogens may be transmitted through the air. There is now considerable information about the constituents of the air flora and its distribution, and the lack of active microbial metabolism during transport removes the problem of biochemical interaction and the interdependence of microbes in the more complex communities in soil and water. There is, however, only limited information on the interactions between micro-organisms, and their microecology, on the aerial surfaces of plants and animals but this important area of research is now beginning to yield useful results.

FURTHER READING

1 Dickinson C.H. & Preece T.F. (Eds.) (1976) *Microbiology of Aerial Plant Surfaces.* Academic Press, London, New York and San Francisco. pp. 669.
Reviews and original papers.

2 Gregory P.H. (1973) *The Microbiology of the atmosphere.* 2nd Edition. Leonard Hill, Aylesbury, England. pp. 377.
Basic text for outside air.

3 Gregory P.H. & Montieth J.L. (Eds.) (1967) *Airborne Microbes.* 17th Symposium of the Society for General Microbiology. Cambridge University Press, Cambridge. pp. 385.
Reviews, still with useful information.

4 Hers J.F.P. & Winkler K.C. (Eds.) (1973) *Airborne Transmission and Airborne Infection.* Oosthoek Publishing, Utrecht, Nederlands. pp. 610.
Particularly useful for inside air.

5 INGOLD C.T. (1971) *Fungal Spores, their Liberation and Dispersal.* Clarendon Press, Oxford. pp. 302.
General text.
6 PREECE T.F. & DICKINSON C.H. (Eds.) (1971) *The Ecology of Leaf-Surface Micro-organisms.* Academic Press, London and New York. pp. 640.
Reviews and original papers.
7 RUINEN J. (1961) The phyllosphere. 1. An ecologically neglected milieu. *Plant and Soil* **15**, 81–109.
The first major review of the subject.
8 SKINNER F.A. & CARR J.G. (Eds.) (1974) *The Normal Microbial Flora of Man.* Symposium Series No. 3, Society for Applied Bacteriology. Academic Press, London and New York. pp. 264.
Particularly useful for medically important organisms.

9 Conclusions

The fact that this is a book on microbial ecology is a reflection of the unfortunate separation of micro-organisms from the rest of ecology. Microbes are an integral part of ecosystems, and may be important in terms of both biomass and activity. There is no reason why microbes should not be subjected to the same basic concepts of community structure, population dynamics and quantitative study, backed by mathematical and statistical techniques, that are applied to larger organisms. Microbiology has problems with methodology but this should be a spur to developing improved methods, rather than an excuse for dealing with microbial ecology in a less rigorous way than is normal for other branches of the subject.

The different areas of microbiology have received very different amounts of study: there are many examples that should be obvious from the previous chapters but protozoology in particular seems to have been neglected in almost all habitats, while algae have possibly been overemphasized in aquatic environments. Bacteria have suffered from the complexity and the difficulty of classifying them, and from the lack of clear species concepts. This leads to vague descriptions of communities, but it is difficult to see the solution to this problem other than a lot of hard work on classifying the bacteria found. Numerical taxonomy may be a long term solution, provided that the 'clusters' of related isolates can be so well defined that a similar group of organisms can be identified in situations other than that for which the particular cluster analysis was done. This returns to the point made in early chapters: descriptive microbial ecology takes more time and work to achieve a similar degree of precision to that obtainable relatively easily with higher plants and animals. Microbial ecologists are in the position that general ecologists were half a century ago: there have been a few uncoordinated collecting trips into particular habitats, but there is little detailed knowledge for most habitats of what species occur where. However physiological investigations on microbes are often easier than with higher plants because of the convenience with which they can be cultured. It is likely that the physiological interactions of micro-organisms in natural environments may be partly understood before full descriptive data are available.

A great deal is known about the ecology of a very small number of micro-organisms that happen to be pathogens of plants and animals. Plant and animal pathology and medicine have, rightly, received much attention. However the overwhelming majority of micro-organisms, which are harmless or beneficial, remain rather neglected. It is hoped that *ad hoc* solutions to particular problems in agriculture, water management etc. will continue to be found, but much more basic information on micro-organisms in ecosystems, whether natural or not, is required before such solutions can be developed on a rational basis.

Micro-organism index

Acanthamoeba 72
Achromobacter 28, 39, 112
Achlya 116, 122
Achnanthes 60, 100, 113, 115, 121
Actinomyces 140
Actinophrys 115
Actinoplanes 115
Agaricus 73
Agrobacterium 80
Alternaria 72, 130, 133, 135, 136
Amanita 87
Amphicypellus 121
Amphidinium 102
Amphiprora 102
Amphora 100, 102, 113, 121
Anabeana 19, 41, 99, 100, 106, 110, 117
Anguillospora 117
Ankistrodesmus 100, 106
Aphanizomenon 41
Aphanothece 115
Arcella 100, 111
Articulospora 117
Arthrobacter 71, 101
Aspergillus 72, 75, 117, 138
Aspicilia 74
Asterionella 106, 107, 108, 121, 124
Aureobasidium 76, 129, 134, 136, 137
Azotobacter 3, 41, 43, 44, 47, 48, 72, 80, 82, 83, 135

Bacillus 25, 39, 41, 72, 75, 80, 89, 101, 117, 133, 134, 139, 140
Bangia 102
Bdellovibrio 25, 102
Beggiatoa 53, 55, 121, 122
Beijerinckia 41, 43, 135
Boletus 87
Botrytis 72
Bulbochaete 102, 121

Caloneis 100, 117
Calothrix 100, 112, 113, 117
Caloplaca 74
Candida 28, 105, 116, 139
Caulobacter 111, 112, 121
Cenococcum 87
Cephalosporium 105
Ceratium 99, 100, 102, 110, 121
Cercomonas 72
Chaetoceros 99, 102
Chaetomium 25, 26
Chaetophora 100
Chamaesiphon 100
Characium 100
Chilodonella 122
Chlamydomonas 60, 73, 99, 100, 107, 133
Chlorella 3, 99, 100, 106, 107, 114, 119, 133
Chlorobium 53, 57, 110, 119
Chlorococcum 19, 73, 107
Chlorflexus 115
Chromobacterium 43
Chromatium 41, 53, 55, 105, 110
Chroococcus 106, 138
Chromulina 99
Chrysococcus 102
Cladophora 112, 120, 121
Cladosporium 28, 72, 105, 130, 131, 133, 136
Clavariopsis 116, 117
Closterium 100
Clostridium 25, 41, 44, 46, 101, 119, 125, 139
Coccolithus 106
Cocconeis 100, 102, 111, 113, 121
Codonella 100
Coelosphaerium 99
Coleps 122
Colpidium 100, 122, 124
Colpoda 72
Coniosporium 76
Corynebacterium 80, 134
Coscinodiscus 99, 100, 102
Cosmarium 99, 106
Crenothrix 115
Cryptococcus 105, 134
Cryptomonas 106, 107
Cyclidium 115
Cyclotella 100, 121
Cylindrocarpon 82, 83
Cymatopleura 100
Cymbella 100, 102, 121
Cytophaga 25, 75, 112, 121

Debaryomyces 105
Denitrobacter 40
Dermocarpa 100
Desmazierella 76
Desulfovibrio 28, 37, 41, 55–57, 95, 110, 118
Diatoma 107, 124
Dictyosphaerium 108
Difflugia 100

Dinobryon 99, 100
Dinophysis 99
Diploneis 102
Disematostoma 67

Ectocarpus 102
Endogone 85
Enteromorpha 117
Epicoccum 136
Epithemia 121
Erythrotrichia 102
Erwinia 134
Erisyphe 130
Escherichia 101, 125
Eucampia 102
Eucocconeis 115
Eudorina 120
Euglena 60, 67, 99, 100, 115, 121
Eunotia 115, 138
Euplotes 100, 114
Exuviella 99

Flavobacterium 28, 101, 112, 119, 121, 134
Fragilaria 99, 100, 102, 106, 113, 124
Frankia 42
Frontonia 114
Frustulia 60, 115
Fusarium 72, 80, 82, 87, 117
Fusicoccum 76

Gaeumannomyces 73, 89
Gallionella 60, 115
Gemellicystis 108
Glaucoma 122
Gliocladium 83
Gliomastix 82, 83
Globigerina 100
Gloeocapsa 19, 41, 74, 112
Gloeocystis 74
Gloeotrichia 106
Gomphonema 100, 111, 120, 121
Gonyaulax 100, 106, 110
Grammatophora 102
Gymnodinium 99, 100, 109, 113

Halosphaeria 112
Hantzschia 73, 102, 113, 137
Hartmanella 72
Helicoma
Herpesvirus 139
Heteromita 72
Holopedia 102
Hormidium 73
Hyalodiscus 115
Hydrogenomonas 31
Hymenomonas 102
Hyphomicrobium 59, 111, 112, 121

Keratinomyces 24
Klebsiella 41, 43

Labrynthula 121
Lactarius 72, 87
Lactobacillus 134
Lampropedia 107
Leptomitus 124
Leptopharynx 75
Leptothrix 60, 115
Leucosporidium 105
Leucothrix 121
Licmophora 102
Lithoderma 111
Lophodermella 134
Lophodermium 76
Lulworthia 117
Lyngbya 99, 100, 112, 113, 118, 119, 121, 124

Mallomonas 99, 100
Mastigocladus 115
Mastogloia 102
Melosira 99, 100, 106
Meridion 124
Merismopedia 100
Methanobacillus 21
Methanomonas 21
Metopus 95
Metschnikowia 103
Micrococcus 39, 101, 133
Microcoleus 117
Microcystis 99, 110
Micromonospora 115
Microsporium 139
Mortierella 25, 72
Mucor 72, 75
Mycobacterium 25, 41

Naegleria 72
Nassula 115
Navicula 100, 102, 112, 117, 118, 121, 137
Nitrobacter 37–39, 46, 47, 72
Nitrosococcus 38
Nitrosocystis 38
Nitrosomonas 37–39, 46, 47, 72
Nitzschia 99, 100, 102, 112, 113, 117, 121, 124
Nocardia 115
Nostoc 19, 41, 73, 74, 112–115, 117, 137, 138

Ochromonas 107
Oedogonium 73, 100, 112, 121
Oikomonas 72
Olpidium 73
Oocystis 107
Opephora 100, 102, 113
Ophiocytium 100
Oscillatoria 99, 100, 106, 121, 133
Oxytricha 60

Paecilomyces 82
Palmogloea 19, 74
Pandorina 100, 107
Paramecium 3, 114, 122

Pediastrum 106
Pedomicrobium 59
Pelomyxa 95
Penicillium 24, 25, 72, 73–75, 83, 105, 117, 129, 138
Peridinium 99, 102, 106, 107, 110
Phacus, 99, 115
Phlyctidium 121
Phormidium 117, 124
Photobacterium 122
Phyllobacterium 43
Phymatotrichum 88
Phytophthora 87, 130, 132
Physcia 74
Picia 117
Pinnularia 100, 137
Pityrosporum 139
Platymonas 113
Plectonema 41
Pleurocapsa 102
Pleurosigma 102, 117
Poria 26
Porolithon 113
Porphyridium 119
Porphyrosiphon 74
Prasiola 69, 112
Propionibacterium 140
Prorodon
Proteus 101
Pseudanabeana 138
Pseudomonas 3, 25, 28, 37, 39, 72, 80, 89, 101, 110–112, 119, 121, 134
Pseudospora 108
Puccinia 131
Pythium 116, 117

Radiococcus 108
Raphoneis 102, 113
Remanella 113
Rhizobium 37, 43, 45–47, 72, 73, 80, 83
Rhizoctonia 80, 83, 85
Rhizophidium 121
Rhizopus 72, 75
Rhizosolenia 102
Rhodomonas 106, 107
Rhodopseudomonas 19, 41, 110, 119
Rhodospirillum 41
Rhodosporidium 105
Rivularia 102, 112, 117
Russula 72, 87

Saccharomyces 3
Salmonella 125
Saprolegnia 116, 117, 122
Sarcina 101
Sceletonema 99, 102, 106
Scenedesmus 99, 100, 107, 109, 121
Schizothrix 74, 112, 119
Scytonema 74, 112
Serpula 26
Shigella 125
Sphaerocystis 108
Sphaerotilus 59, 60, 111, 121, 124, 125
Spiretta 67
Spirillum 82, 101
Spirofilopsis 67
Spirogyra 112, 115, 124
Spirostomum 113
Spirulina 100
Sporobolomyces 105, 130, 134
Stachybotrys 25, 26
Staphylococcus 139, 140
Staurastrum 99, 100, 108
Stentor 114
Stephanodiscus 100, 106, 107, 121
Sympodiella 74
Stephanopyxis 106
Stichococcus 73, 133
Streptococcus 101, 140
Streptomyces 24, 25, 72, 75, 88, 115
Surirella 100, 118
Synechococcus 115, 119
Synedra 102, 120, 121

Tabellaria 100, 106
Tetracladium 116, 117
Tetracystis 133
Thalassionema 102
Thalassiosira 99, 106
Thiobacillus 39, 53–55, 57, 60
Thiocapsa 19
Thiothrix 53–55, 110
Torulopsis 134, 139
Tolypothrix 112
Trachelocerca 113
Trachelomonas 100, 107
Trentepohlia 134
Trichoderma 25, 75, 76, 82, 83
Trichodesmium 99
Trichodina 122
Tricholoma 87
Trichophyton 139
Trichosporon 117
Tricladium 117
Tropidoneis 102

Ulothrix 100, 111, 117, 124

Varicosporium 116, 117
Vaucheria 73, 117
Verrucaria 74, 112
Verticillium 83
Vibrio 101, 112, 121

Woloszynskia 107

Xanthomonas 43, 134
Xanthoria 74

Zygnema 115
Zygogonium 73, 74

Subject index

Adsorption 22, 36, 47, 51, 65–66, 97
Aeration 1, 4
  carbon cycle 21, 23, 25
  nitrogen cycle 39, 41, 46, 48
  soil 64, 68, 90
  sulphur cycle 55–56, 118
  water 36, 93–95
Agricultural practices
  air 129
  carbon cycle 32
  nitrogen cycle 46–48
  soil 59, 89–90
Air
  chemistry and physics 128
  microbiology 128–133, 138–139
Animal surfaces 139–140
Antibiotics 4, 79, 97
Arthropods 12, 73, 75, 77
ATP in oceans 5
Autochthonous flora 69

Biocides 29–32
Biological control 87–89, 136
Blooms 105–109, 113, 115, 121

Carbon dioxide
  air 32
  soil 19, 64, 79
  water 21, 93–94
Carbon cycle 17–33
Carbonate/bicarbonate equilibrium 19, 21, 23, 61, 94–95, 114–115
Cellulose decay 25
Chitin decay 25
Competition 15, 43
  animal surfaces 140
  leaves 136
  soil 73–78
  water 100, 118
Consumers 9, 19, 108
Coral reefs 112–113
Cultural methods, 3, 16, 30

Decay prevention 28
Denitrification 39–40, 48, 83
Detritus 9–12, 22, 117, 123
Dilution plates 4
Direct examination 3

Diurnal changes
  air flora 130–131
  epipelon 96, 118
  epipsammon 113
  plankton 96, 110

Eh 53–54, 56, 60, 77
Energy flow 9, 17
Epilithon 111–113
Epipelon 54, 114–119
Epiphyton 119–122
Epipsammon 113–114
Epizoon 122
Euryhaline 95, 103
Eutrophic 97, 99, 106
Eutrophication 59, 122–126

Faeces in nutrient cycling 12, 19, 75, 77, 109
Fertilizers 46–47
Food chains 9–12, 77, 122
Fungicides 29–32, 136
Fungistasis 69

Generation times 16
Growth forms in soil 66, 83

Herbicides 29–32, 47
Hot springs 115

Iron 59–60

Leaf nodules 43
Leaf surface 132–138
Lichens 41, 74, 112
Light 1, 23, 55, 96, 105, 107, 109, 118–119
Lignin decay 26
Lipid decay 24

Manganese 59–60
Metabolic activity measurement 3, 5, 82
Metal corrosion 56
Methane 21
Methods 1–7
Models of ecosystems 7

Microhabitat 2, 25, 54, 66, 68, 77, 133
Minerals 14, 23
Mucopeptide 25
Mycorrhiza 59, 72, 85–87

Nanoplankton 99
Neuston 111, 129
Nitrate reduction 39–40
Nitrification 37–39, 46, 47, 72
Nitrogen 23
  cycle 34–39
  fixation 34, 40–46, 49, 82, 112, 113, 117
  soil 36–38, 46–48
  water 22, 34–36, 48, 59, 124
Nutrient cycles 13, 119
  carbon 17–33
  iron 59–60
  nitrogen 34–49
  phosphorus 56–59
  silicon 60–61
  sulphur 51–56

Oligotrophic 97, 99, 115
Organic matter
  decomposition 14, 24–29, 48
  soil 64, 75, 79
  water 22, 97
Oxygen 14, 15
  soil 64, 79
  water 21, 93–94, 115–116, 125

Paint decay 27
Parasitic bacteria 25, 102, 139–140
  fungi 12, 73, 87–89, 103, 108, 121, 131–132, 134, 136, 138
  protozoa 108
  viruses 73, 102
Pesticide breakdown 29–32
Petroleum decay 28
pH 1, 54
  carbon cycle 23
  iron 59–60
  nitrification 47
  soil 54, 64, 68, 77, 87
  sulphur cycle 51, 53–56
  water 94–95
Phosphorus
  soil 57–58, 85, 87
  water 57, 124
Phyllosphere 133–138
Plankton 9, 36, 57, 99–111, 129
Plant diseases
  biological control 87–89, 136
  fertilizer effect 48
  leaves 134–136
  spread 73, 131–132
Plant hormones 46, 47, 79, 83
Plant nutrient uptake 59, 87
Plastics 27
Pollution 32, 57, 122–126
  atmospheric 137
  biocides 29–32
  sewage 29, 32, 46, 100
Predation 13, 73–78, 108
Primary producers 9, 18–19, 21, 98–99, 111
Productivity 9, 14, 17, 18, 21, 93, 96, 100, 124
Protein decay 24

Rhizosphere 78–87
Root nodules 41–44
Root surface 78–87
Rubber decay 27

Salt marshes 117–118
Seasonal changes in micro-organisms
  air 129–130
  leaves 135–136
  soil 69
  water 105–106, 112–116
Seral changes 15–16, 73–75
Sewage 29, 32, 35, 46, 59, 101, 124
Silicon 60–61, 74, 106
Soil
  chemistry 63
  formation 73–75
  micro-organisms 66–73, 83
  structure 64–65
Stenohaline 96, 103
Stratification 92–94
Substrate successions 75–77, 117
Succession 15
  leaves 135–136
  soil 73–78
  water 106, 111–112
Sulphur 51–56
  bacteria 28, 41, 55–56, 60, 69, 110, 115, 118
Symbiosis 15, 37, 41–46, 59, 72, 74, 76, 80, 83, 85, 100, 113, 114

Temperature 1, 4
  carbon cycle 23
  nitrogen cycle 48
  soil 64, 69
  water 48, 92–94, 115
Thermocline 92–94, 105
Trophic levels 12–13

Virus 73, 102, 132

Water
  currents 96–97, 111
  chemistry and physics 92
  pressure 97–98
  salinity 96

Zymogenous flora 69, 75, 89